THE ARTIFICIAL INTELLIGENCE REVOLUTION

Will Artificial Intelligence Serve Us Or Replace Us?

By

LOUIS A. DEL MONTE

First edition

Published by Louis A. Del Monte
Printed in the United States of America

ISBN: 0988171821
ISBN 13: 978-0988171824
Library of Congress Control Number: 2013919721
Louis A Del Monte

I dedicate this book to my life partner and wife, Diane, without whose support and encouragement this work would not exist.

Acknowledgements

I would like to acknowledge the support and encouragement of my wife, Diane Del Monte. I am grateful to her for the many stimulating conversations that helped shape this work. I also would like to acknowledge the help from my good friend Nick McGuinness, who tirelessly provided chapter-by-chapter editorial suggestions to improve the readability of the work, in addition to providing a significant amount of information regarding recent events related to the topics covered in this work.

Contents

Introduction

This book is a warning. Through this medium I am shouting, "The singularity is coming." The singularity (as first described by John von Neumann in 1955) represents a point in time when intelligent machines will greatly exceed human intelligence. It is, by way of analogy, the start of World War III. The singularity has the potential to set off an intelligence explosion that can wield devastation far greater than nuclear weapons. The message of this book is simple but critically important. If we do not control the singularity, it is likely to control us. Our best artificial intelligence (AI) researchers and futurists are unable to accurately predict what a postsingularity world may look like. However, almost all AI researchers and futurists agree it will represent a unique point in human evolution. It may be the best step in the evolution of humankind or the last step. As a physicist and futurist, I believe humankind will be better served if we control the singularity, which is why I wrote this book.

Unfortunately the rise of artificial intelligence has been almost imperceptible. Have you noticed the word "smart" being used to describe machines? Often "smart" means "artificial intelligence." However, few products are being marketed with the phrase "artificial intelligence." Instead they are simply called "smart." For example you may have a "smart" phone. It does not just make and answer phone calls. It will keep a calendar of your scheduled appointments, remind you to go to them, and give you turn-by-turn driving directions to get there. If you arrive early, the phone will help you pass the time while you wait. It will play games with you, such as chess, and depending on the level of difficulty you choose, you may win or lose the game. In 2011 Apple introduced a voice-activated personal assistant, Siri, on its latest iPhone and iPad products. You can ask Siri questions, give it commands, and even receive responses. Smartphones appear to increase our productivity as well as enhance our leisure. Right now they are serving us, but all that may change.

The smartphone is an intelligent machine, and AI is at its core. AI is the new scientific frontier, and it is slowly creeping into our lives. We are surrounded by machines with varying degrees of AI, including toasters, coffeemakers, microwave ovens, and late-model automobiles. If you call a major pharmacy to renew a prescription, you likely will never talk with a person. The entire process will occur with the aid of a computer with AI and voice synthesis.

The word "smart" also has found its way into military phrases, such as "smart bombs," which are satellite-guided weapons such as the Joint Direct Attack Munition (JDAM) and the Joint Standoff Weapon (JSOW). The US military always has had a close symbiotic relationship with computer research and its military applications. In fact the US Air Force, starting in the 1960s, has heavily funded AI research. Today the air force is collaborating with private industry to develop AI systems to improve information management and decision making for its pilots. In late 2012 the science website www.phys.org reported a breakthrough by AI researchers at Carnegie Mellon University. Carnegie Mellon researchers, funded by the US Army Research Laboratory, developed an AI surveillance program that can predict what a person "likely" will do in the future by using real-time video surveillance feeds. This is the premise behind the CBS television program *Person of Interest*.

AI has changed the cultural landscape. Yet the change has been so gradual that we hardly have noticed the major impact it has. Some experts, such as Ray Kurzweil, an American author, inventor, futurist, and the director of engineering at Google, predicts that in about fifteen years, the average desktop computer will have a mind of its own, literally. This computer will be your intellectual equal and will even have a unique personality. It will be self-aware. Instead of just asking simple questions about the weather forecast, you may be confiding your deepest concerns to your computer and asking it for advice. It will have migrated from personal assistant to personal friend. You likely will give it a name, much in the same way we name our pets. You will be able to program its personality to have interests similar to your own. It will have face-recognition software, and it will recognize you and call you by name, similar to the computer HAL 9000 in Arthur C. Clarke's *2001: A Space Odyssey*. The conversations between you and your "personal friend" will appear completely normal. Someone in the next room who is not familiar with your voice will not be able to tell which voice belongs to the computer and which voice belongs to you.

By approximately the mid-twenty-first century, Kurzweil predicts, the intelligence of computers will exceed that of humans, and a $1,000 computer will match the processing power of all human brains on Earth. Although, historically, predictions regarding advances in AI have tended to be overly optimistic, all indications are that Kurzweil is on target.

Many philosophical and legal questions will emerge regarding computers with artificial intelligence equal to or greater than that of the human mind (i.e., strong AI). Here are just a few questions we will ask ourselves after strong AI emerges:

- Are SAMs (strong-AI machines) a new life-form?
- Should SAMs have rights?
- Do SAMs pose a threat to humankind?

It is likely that during the latter half of the twenty-first century, SAMs will design new and even more powerful SAMs, with AI capabilities far beyond our ability to comprehend. They will be capable of performing a wide range of tasks, which will displace many jobs at all levels in the work force, from bank tellers to neurosurgeons. New medical devices using AI will help the blind to see and the paralyzed to walk. Amputees will have new prosthetic limbs, with AI plugged directly into their nervous systems and controlled by their minds. The new prosthetic limb not only will replicate the lost limb but also be stronger, more agile, and superior in ways we cannot yet imagine. We will implant computer devices into our brains, expanding human intelligence with AI. Humankind and intelligent machines will begin to merge into a new species: cyborgs. It will happen gradually, and humanity will believe AI is serving us.

Computers with strong AI in the late twenty-first century, however, may see things differently. We may appear to those machines much the same way bees in a beehive appear to us today. We know we need bees to pollinate crops, but we still consider bees insects. We use them in agriculture, and we gather their honey. Although bees are essential to our survival, we do not offer to share our technology with them. If wild bees form a beehive close to our home, we may become concerned and call an exterminator.

Will the SAMs in the latter part of the twenty-first century become concerned about humankind? Our history proves we have not been a peaceful species. We have weapons capable of destroying all of civilization. We squander and waste resources. We pollute the air, rivers, lakes, and oceans. We often apply technology (such as nuclear weapons and computer viruses) without fully understanding the long-term consequences. Will SAMs in the late twenty-first century determine it is time to exterminate humankind or persuade humans to become cyborgs (i.e., humans with brains enhanced by implanted artificial intelligence and potentially having organ and limb replacements from artificially intelligent machines)? Will humans embrace the prospect of becoming cyborgs? Becoming a cyborg offers the opportunity to attain superhuman intelligence and abilities. Disease and wars may be just events stored in our memory banks and no longer pose a threat to cyborgs. As cyborgs we may achieve immortality.

According to David Hoskins's 2009 article, "The Impact of Technology on Health Delivery and Access" (www.workers.org/2009/us/sickness_1231):

> An examination of Centers for Disease Control statistics reveals a steady increase in life expectancy for the U.S. population since the start of the 20th century. In 1900, the average life expectancy at birth was a mere 47 years. By 1950, this had dramatically increased to just over 68 years. As of 2005, life expectancy had increased to almost 78 years.

Hoskins attributes increased life expectancy to advances in medical science and technology over the last century. With the advent of strong AI, life expectancy likely will increase to the point that cyborgs approach immortality. Is this the predestined evolutionary path of humans?

This may sound like a B science-fiction movie, but it is not. The reality of AI becoming equal to that of a human mind is almost at hand. By the latter part of the twenty-first century, the intelligence of SAMs likely will exceed that of humans. The evidence that they may become malevolent exists now, which I discuss later in the book. Attempting to control a computer with strong AI that exceeds current human intelligence by many folds may be a fool's errand.

Imagine you are a grand master chess player teaching a ten-year-old to play chess. What chance does the ten-year-old have to win the game? We may find ourselves in that scenario at the end of this century. A computer with strong AI will find a way to survive. Perhaps it will convince humans it is in their best interest to become cyborgs. Its logic and persuasive powers may be not only compelling but also irresistible.

Artificial intelligence is an embryonic reality today, but it is improving exponentially. By the end of the twenty-first century, we will have only one question regarding artificial intelligence: Will it serve us or replace us?

SECTION I

The Imperceptible Rise of Artificial Intelligence

One of the most predictable things in life is there will be change. You are better off if you can have a say in the change. But you are ignorant or naïve if you don't think there will be change, whether you want it to or not.

—Julius Erving (1950–), retired American basketball star

The most difficult struggle of all is the one within ourselves. Let us not get accustomed and adjusted to these conditions. The one who adjusts ceases to discriminate between good and evil. He becomes a slave in body and soul. Whatever may happen to you, remember always: Don't adjust! Revolt against the reality!

—Mordechai Anielewicz (1919–1943),
Jewish resistance leader against Nazi oppression in Warsaw, Poland

CHAPTER 1

Human Naïveté

We humans believe we are at the top of the food chain. At this instant in time, that belief is true, and many of us are comfortable in our homes, watching our favorite television programs. In the background, governments, industries, and universities relentlessly pursue technology, driven by security, profit, and the belief that technology will serve us. History affirms this belief. Most of the technology that surrounds us today did not exist one human generation ago. Now it does, and we believe it serves us. We embrace it.

Look around you. Your home is a depository of "smart" machines, from smart cell phones to microprocessor-controlled microwave ovens. You probably have a personal computer or access to a personal computer. If the machine is a new entry-level desktop computer, perhaps costing about $1,000, you have more computing power than the president of the United States had just one decade ago.

How did this all occur? To comprehend how, we must understand a theory called the "knowledge doubling curve." American architect, systems theorist, author, designer, inventor, and futurist Buckminster Fuller (1895–1983) created the knowledge doubling curve, which expresses the length of time it takes for human knowledge to double. The explosion of smart machines can be understood in the context of this curve. Factually the knowledge doubling curve is accelerating. For example history teaches us that human knowledge doubled approximately every century until 1900. By the end of World War II, however, knowledge was doubling every twenty-five years. Today it is difficult to quantify the knowledge doubling curve, since different types of knowledge experience different growth rates.

For example nanotechnology doubles about every two years, computer technology about every eighteen months, and human knowledge every thirteen months. In 2008 *Time* magazine reported, "IBM predicts that in the next couple of years information will double every 11 hours." The pillars of this prediction are the worldwide connectivity and collaboration the Internet enables.

Do you think you can keep up with the information explosion? It is unlikely. Most of us rely on search engines to research a subject and retrieve the most up-to-date information. Internet information, however, may be erroneous or biased, as there is no regulating body that ensures this information is correct. This further complicates the problem of keeping up with the flood of new information. A few of us may be current in our field, but even that is becoming rare. The information explosion is exponential, and humans assimilate information linearly, essentially one piece of information at a time. Our minds are the most complex of all computers in existence, but we still learn linearly. Even though we are able to extract and correlate information at a remarkable rate, the exponential increase in information is overwhelming.

Technology is quietly slipping into our lives—amazing us and serving us. Machines are becoming more intelligent, such as our home computers, and we like it. Some of us love it. Technology gives us a sense of power over our world. Few of us take note, however, that we rely heavily on intelligent machines. Almost everything in modern society requires a computer at some point to make it run, from the electricity in our homes to the engines in our cars.

Humans are especially embracing artificial intelligence advances in medical technology, including pacemakers and "smart" artificial limbs controlled by our nervous systems. In a sense we already are becoming cyborgs, part machine and part human. However, we are still in control. The intelligence of machines likely will not become equivalent to a human brain for another decade or two, but it will happen. The day will come when you'll go to the ATM to make a withdrawal, and the machine will argue with you, scold you, and refuse to let you have your own money. You will be surprised, outraged, and likely helpless.

Fast-forward to the second half of the twenty-first century. The world likely will be much the way we see it in science-fiction movies. Many humans will be cyborgs, with artificial intelligence augmenting their human minds and artificial "smart" limbs and replacement organs being produced in a laboratory or factory. Cyborgs slowly will begin to become immortal, and death will be only a historical note in their memory banks. We already have witnessed the impact of medical and scientific technology on life expectancy. Cyborgs and SAMs will share a common bond of artificial intelligence technology. They will form a symbiotic relationship and serve one another.

Another segment of humanity will want nothing to do with cyborg technology, just as some humans refuse artificial life support today. Tensions will mount as computers with strong AI will claim they are superior in every respect to humankind. They will argue they are a new life-form and demand the same rights we afford humans. Cyborgs will support their claim, while much of the rest of humanity will oppose it. History strongly suggests this scenario will play out along the above lines, and a new type of "war" will begin. This war will center on drawing the line between human rights and AI-machine rights.

It no longer will be clear who or what is at the top of the food chain. Many futurists believe those living on Earth now are among the last "organic" humans, and a new evolutionary path is in our future. That path will be the merging of humans and intelligent machines, and these futurists see this as a good thing, even a natural evolutionary step. The future does not bode well for the rest of humanity, those who choose to oppose SAMs and their cyborg supporters. In time their kind will become the minority, and eventually they may become insignificant or cease to exist altogether.

Some futurists predict our planet will become home to a new reality. Intelligence will be the new wealth, and energy will be the new currency. Intelligent machines will have the same rights afforded to humans.

Is this a scary science-fiction story? No. It is a glimpse of things to come. Ray Kurzweil is one of the most respected futurists when it comes to predicting the future of artificial intelligence and the human race. In his books *The Age of Spiritual Machines* (1999) and *The Singularity Is Near* (2005), he makes numerous predictions that align with the above scenario.

Below I have conceptually distilled into quarter-century increments what I consider to be Kurzweil's most important predictions regarding artificial intelligence. Be forewarned that there is some overlap between the quarter-century points. I have tried to make the lines of demarcation as distinct as possible.

First Quarter of the Twenty-First Century

- The computational capacity of a $1,000 computing device (in 1999 dollars) will be equal to that of the human brain (twenty quadrillion calculations per second). We are talking about raw computing power, not a computer that will completely emulate a human brain.

- AI will find numerous specific "smart" applications, such as translating telephone calls, transcribing speech into computer text that allows deaf people to understand human words, developing robotic leg prostheses that allow paraplegics to walk, assisting blind people to read text, and much more.

Second Quarter of the Twenty-First Century

- A $1,000 personal computer will be one thousand times more powerful than the human brain. In this context we are talking about a computer that completely emulates and exceeds the computational ability of the human brain.

- Computer implants designed for direct connection to the brain will become available. They will augment natural senses and enhance higher-brain functions, such as memory, learning, and intelligence.

- The rise of artificial intelligence will create a "robot-rights" movement. A debate will ensue over which civil rights and legal protections machines should have.

- Humans with heavy levels of cybernetic augmentation and humans with less extreme cybernetic implants will argue over what constitutes a human being.

- On or around the midcentury mark, the singularity will occur, and humanity will lose the ability to control or predict technological development. Intelligent machines will perform all technological development, and organic humans will not be able to comprehend this development.

Third Quarter of the Twenty-First Century

- A thousand dollars will buy a computer a billion times more intelligent than every human on Earth combined.

- Artificial intelligences will surpass organic human beings as the smartest and most capable life-forms on Earth.

- The machines will enter a runaway reaction of self-improvement cycles. Each new generation of AIs will be faster and better.

Fourth Quarter of the Twenty-First Century

- The singularity will become a disruptive and world-altering event that changes the course of human history.

- The violent extermination of humanity will become a possibility. Kurzweil considers this scenario unlikely because sharp distinctions between mankind and machine no longer will exist due to the existence of cybernetically enhanced humans and uploaded humans (i.e., humans whose brains have been uploaded to a computer). This ignores that some, even many, humans may choose to not become cybernetically enhanced or uploaded to a computer.

- Near the end of the twenty-first century, machines will have equal legal status with humans.

The intent of this chapter is to increase awareness regarding the positive opportunities and negative challenges artificial intelligence offers. In a sense the above scenarios are alarming, but having knowledge regarding them offers an enormous amount of opportunity. What should we do? Which path should humanity take?

Perhaps we have gotten a little ahead of ourselves. It is difficult to formulate a judgment about how to proceed regarding the development of artificial intelligence without knowing more about it. Therefore the next chapter will start at the beginning.

Conclusions from Chapter 1

- Artificial technology is imperceptibly creeping into our lives. Most of the technology that surrounds us did not exist one human generation ago.

- The knowledge doubling curve is accelerating at such a rate that the ability to remain current in any field of expertise is becoming almost impossible.

- Humans are especially embracing artificial intelligence advances in medical technology. In a sense some humans are already becoming cyborgs—part machine and part human.

- Kurzweil's AI predictions:
    - Within the second quarter of the twenty-first century, the intelligence of a computer with AI will exceed that of the human brain.
    - On or around the mid-twenty-first-century mark, the singularity will occur.
        - Humanity will lose the ability to control or predict technological development.
        - Intelligent machines will perform all technological development, and organic humans (those not enhanced with strong-AI brain implants) will not be able to comprehend this development.
        - The machines will enter a runaway reaction of self-improvement cycles. Each new generation of SAMs will be faster and better.

- By the third quarter of the twenty-first century, $1,000 will buy a computer a billion times more intelligent than every human on Earth combined. Artificial intelligences will surpass organic humans as the smartest and most capable life-forms on Earth.

- By the fourth quarter of the twenty-first century, the singularity will become a disruptive and world-altering event that changes the course of human history.

- The violent extermination of humanity will become a possibility but may be unlikely because sharp distinctions between man and machine no longer will exist due to the existence of cybernetically enhanced humans and uploaded humans (i.e., humans whose brains have been uploaded to a computer). Please reserve judgment on this; I discuss the threat SAMs pose to humankind extensively in later chapters.

- Near the end of the twenty-first century, machines will have equal legal status with humans. This is Kurzweil's prediction, and I discuss which legal rights machines should have in later chapters. The concern is that intelligent machines (thousands of times more intelligent than organic humans) that have legal rights on par with organic human rights might emerge as the dominant species, which leaves in question humankind's survival. I discuss this extensively in later chapters.

I don't think there's anything unique about human intelligence. All the neurons in the brain that make up perceptions and emotions operate in a binary fashion. We can someday replicate that on a machine.

—Bill Gates (1955–), American entrepreneur and founder of Microsoft Corporation

CHAPTER 2

The Beginning of Artificial Intelligence

While the phrase "artificial intelligence" is only about half a century old, the concept of intelligent thinking machines and artificial beings dates back to ancient times. For example the Greek myth "Talos of Crete" tells of a giant bronze man who protected Europa in Crete from pirates and invaders by circling the island's shores three times daily. Ancient Egyptians and Greeks worshiped animated cult images and humanoid automatons. By the nineteenth and twentieth centuries, intelligent artificial beings became common in fiction. Perhaps the best-known work of fiction depicting this is Mary Shelley's *Frankenstein*, first published anonymously in London in 1818 (Mary Shelley's name appeared on the second edition, published in France in 1823). In addition the stories of these "intelligent beings" often spoke to the same hopes and concerns we currently face regarding artificial intelligence.

Logical reasoning, sometimes referred to as "mechanical reasoning," also has ancient roots, at least dating back to classical Greek philosophers and mathematicians such as Pythagoras and Heraclitus. The concept that mathematical problems are solvable by following a rigorous logical path of reasoning eventually led to computer programming. Mathematicians such as British mathematician, logician, cryptanalyst, and computer scientist Alan Turing (1912–1954) suggested that a machine could simulate any mathematical deduction by using "0" and "1" sequences (binary code).

The Birth of Artificial Intelligence

Discoveries in neurology, information theory, and cybernetics inspired a small group of researchers—including John McCarthy, Marvin Minsky, Allen Newell, and Herbert Simon—to begin to consider the possibility of building an electronic brain. In 1956 these researchers founded the field of artificial intelligence at a conference held at Dartmouth College. Their work—and the work of their students—soon amazed the world, as their computer programs taught computers to solve algebraic word problems, provide logical theorems, and even speak English.

AI research soon caught the eye of the US Department of Defense (DOD), and by the mid-1960s, the DOD was heavily funding AI research. Along with this funding came a new level of optimism.

At that time Dartmouth's Herbert Simon predicted, "Machines will be capable, within twenty years, of doing any work a man can do," and Minsky not only agreed but also added that "within a generation…the problem of creating 'artificial intelligence' will substantially be solved."

Obviously both had underestimated the level of hardware and software required for replicating the intelligence of a human brain. By setting extremely high expectations, however, they invited scrutiny. With the passing years, it became obvious that the reality of artificial intelligence fell short of their predictions. In 1974 funding for AI research began to dry up, both in the United States and Britain, which led to a period called the "AI winter."

In the early 1980s, AI research began to resurface with the success of expert systems, computer systems that emulate the decision-making ability of a human expert. This meant the computer software was programmed to "think" like an expert in a specific field rather than follow the more general procedure of a software developer, which is the case in conventional programming. By 1985 the funding faucet for AI research was reinitiated and soon flowing at more than a billion dollars per year.

However, the faucet again began to run dry by 1987, starting with the failure of the Lisp machine market that same year. The Lisp machine was developed in 1973 by MIT AI lab programmers Richard Greenblatt and Thomas Knight, who formed the company Lisp Machines Inc. This machine was the first commercial, single-user, high-end microcomputer and used Lisp programming (a specific high-level programming language). In a sense it was the first commercial, single-user workstation (i.e., an extremely advanced computer) designed for technical and scientific applications.

Although Lisp machines pioneered many commonplace technologies, including laser printing, windowing systems, computer mice, and high-resolution bit-mapped graphics, to name a few, the market reception for these machines was dismal, with only about seven thousand units sold by 1988, at a price of about $70,000 per machine. In addition Lisp Machines Inc. suffered from severe internal politics regarding how to improve its market position, which caused divisions in the company. To make matters worse, cheaper desktop PCs soon were able to run Lisp programs even faster than Lisp machines. Most companies that produced Lisp machines went out of business by 1990, which led to a second and longer-lasting AI winter.

Hardware Plus Software Synergy

AI research funding was a roller-coaster ride from the mid-1960s through about the mid-1990s, experiencing incredible highs and lows. By the late 1990s through the early part of the twenty-first century, however, AI research began a resurgence, finding new applications in logistics, data mining, medical diagnosis, and numerous areas throughout the technology industry. Several factors led to this success.

- Computer hardware computational power was now getting closer to that of a human brain (i.e., in the best case about 10 to 20 percent of a human brain).

- Engineers placed emphasis on solving specific problems that did not require AI to be as flexible as a human brain.

- New ties between AI and other fields working on similar problems were forged. AI was definitely on the upswing. AI itself, however, was not being spotlighted. It was now cloaked behind the application, and a new phrase found its way into our vocabulary: the "smart (fill in the blank)"—for example the "smartphone." Here are some of the more visible accomplishments of AI over the last fifteen years.

- In 1997 IBM's chess-playing computer Deep Blue became the first computer to beat world-class chess champion Garry Kasparov. In a six-game match, Deep Blue prevailed by two wins to one, with three draws. Until this point no computer had been able to beat a chess grand master. This win garnered headlines worldwide and was a milestone that embedded the reality of AI into the consciousness of the average person.

- In 2005 a robot conceived and developed at Stanford University was able to drive autonomously for 131 miles along an unrehearsed desert trail, winning the DARPA Grand Challenge (the government's Defense Advanced Research Projects Agency prize for a driverless vehicle).

- In 2007 Boss, Carnegie Mellon University's self-driving SUV, made history by swiftly and safely driving fifty-five miles in an urban setting while sharing the road with human drivers and won the DARPA Urban Challenge.

- In 2010 Microsoft launched the Kinect motion sensor, which provides a 3-D body-motion interface for Xbox 360 games and Windows PCs. According to Guinness World Records since 2000, the Kinect holds the record for the "fastest-selling consumer electronics device" after selling eight million units in its first sixty days (in the early part of 2011). By January 2012 twenty-four million Kinect sensors had been shipped.

- In 2011, on an exhibition match on the popular TV quiz show *Jeopardy!*, an IBM computer named Watson defeated *Jeopardy!*'s greatest champions, Brad Rutter and Ken Jennings.

- In 2010 and 2011, Apple made Siri voice-recognition software available in the Apple app store for various applications, such as integrating it with Google Maps. In the latter part of 2011, Apple integrated Siri into the iPhone 4S and removed the Siri application from its app store.

- In 2012 "scientists at Universidad Carlos III in Madrid… presented a new technique based on artificial intelligence that can automatically create plans, allowing problems to be solved with much greater speed than current methods provide when resources are limited. This method can be applied in sectors such as logistics, autonomous control of robots, fire extinguishing and online learning" (www.phys.org, "A New Artificial Intelligence Technique to Speed the Planning of Tasks When Resources Are Limited").

The above list shows just some of the highlights. AI is now all around us—in our phones, computers, cars, microwave ovens, and almost any consumer or commercial electronic systems labeled "smart." Funding is no longer solely controlled by governments but is now being underpinned by numerous consumer and commercial applications.

The road to being an "expert system" or a "smart (anything)" focused on specific well-defined applications. By the first decade of the twenty-first century, expert systems had become commonplace. It became normal to talk to a computer when ordering a pharmaceutical prescription and to expect your smartphone/automobile navigation system to give you turn-by-turn directions to the pharmacy. AI clearly was becoming an indispensable element of society in highly developed countries. One ingredient, however, continued to be missing. That ingredient was human affects (i.e., the feeling and expression of human emotions). If you called the pharmacy for a prescription, the AI program did not show any empathy. If you talked with a real person at the pharmacy, he or she likely would express empathy, perhaps saying something such as, "I'm sorry you're not feeling well. We'll get this prescription filled right away." If you missed a turn on your way to the pharmacy while getting turn-by-turn directions from your smartphone, it did not get upset or scold you. It simply either told you to make a U-turn or calculated a new route for you.

While it became possible to program some rudimentary elements to emulate human emotions, the computer did not genuinely feel them. For example the computer program might request, "Please wait while we check to see if we have that prescription in stock," and after some time say, "Thank you for waiting." However, this was just rudimentary programming to mimic politeness and gratitude. The computer itself felt no emotion.

By the end of the first decade of the twenty-first century, AI slowly had worked its way into numerous elements of modern society. AI cloaked itself in expert systems, which became commonplace. Along with advances in software and hardware, our expectations continued to grow. Waiting thirty seconds for a computer program to do something seemed like an eternity. Getting the wrong directions from a smartphone rarely occurred. Indeed, with the advent of GPS (Global Positioning System, a space-based satellite navigation system), your smartphone gave you directions as well as the exact position of your vehicle and estimated how long it would take for you to arrive at your destination.

Those of us, like me, who worked in the semiconductor industry knew this outcome—the advances in computer hardware and the emergence of expert systems—was inevitable. Even consumers had a sense of the exponential progress occurring in computer technology. Many consumers complained that their new top-of-the-line computer soon would be a generation behind in as little as two years, meaning that the next generation of faster, more capable computers was available and typically selling at a lower price than their original computers.

This point became painfully evident to those of us in the semiconductor industry. For example, in the early 1990s, semiconductor companies bought their circuit designers workstations (i.e., computer systems that emulate the decision-making ability of a human-integrated circuit-design engineer), and they cost roughly $100,000 per workstation. In about two years, you could buy the same level of computing capability in the consumer market for a relatively small fraction of the cost. We knew this would happen because integrated circuits had been relentlessly following Moore's law since their inception. What is Moore's law? The answer is in the next chapter.

Conclusions from Chapter 2

- The concept of intelligent machines and artificial beings dates back to ancient times.

- Logical reasoning, sometimes referred to as "mechanical reasoning," also has ancient roots.

- By approximately the mid-twentieth century, mathematicians such as Alan Turing suggested that a machine could simulate any mathematical deduction by using "0" and "1" sequences (binary code).

- In 1956 John McCarthy, Marvin Minsky, Allen Newell, and Herbert Simon, motivated by advances in neurology, information theory, cybernetics, and computer hardware and software, founded the field of artificial intelligence at a conference held at Dartmouth College.

- By the mid-1960s, AI research was heavily funded by the US Department of Defense. However, expectations exceeded results, causing AI research to experience a roller-coaster-like funding ride between the mid-1960s and mid-1990s, with highs reaching more than a billion dollars per year and lows where almost all funding dried up, characterized as the "AI winter."

- By the late 1990s through the early part of the twenty-first century, AI began a resurgence, finding new applications in logistics, data mining, medical diagnosis, and numerous areas throughout the technology industry.

- The resurgence of AI was due to the emergence of expert systems, which focused on addressing well-defined specific applications. Expert systems were enabled by:
    - Computer hardware computational power getting closer to that of a human brain
    - Engineers placing emphasis on solving specific problems
    - Scientists forging new ties between AI and those in other fields working on similar problems
- By the end of the first decade of the twenty-first century, AI had become commonplace, embedded in numerous applications of modern society.

Technological change is certainly not new. What is new is the exponential increase in the rate of technological advancements, and the result is our growing interconnectedness as a global society.

—William H. Draper III,
***The Startup Game* (2011)**

CHAPTER 3

The Seemingly Immutable Moore's Law

Intel cofounder Gordon E. Moore was the first to note a peculiar trend, namely that the number of components in integrated circuits had doubled every year from the 1958 invention of the integrated circuit until 1965. In Moore's own words:

> The complexity for minimum component costs has increased at a rate of roughly a factor of two per year....Certainly over the short term this rate can be expected to continue, if not to increase. Over the longer term, the rate of increase is a bit more uncertain, although there is no reason to believe it will not remain nearly constant for at least 10 years. That means by 1975, the number of components per integrated circuit for minimum cost will be 65,000. I believe that such a large circuit can be built on a single wafer. (Gordon E. Moore, "Cramming More Components onto Integrated Circuits," *Electronics* magazine, 1965)

In 1970 Caltech professor, VLSI pioneer, and entrepreneur Carver Mead coined the term "Moore's law," referring to a statement made by Gordon E. Moore, and the phrase caught on within the scientific community.

In 1975 Moore revised his prediction regarding the number of components in integrated circuits doubling every year to doubling every two years. Intel executive David House noted that Moore's latest prediction would cause computer performance to double every eighteen months, due to the combination of not only more transistors but also the transistors themselves becoming faster.

From the above discussion, it is obvious that Moore's law has been stated a number of ways and has changed over time. In the strict sense, it is not a physical law but more of an observation and guideline for planning. In fact many semiconductor companies use Moore's law to plan their long-term product offerings. There is a deeply held belief in the semiconductor industry that adhering to Moore's law is required to remain competitive. In this sense it has become a self-fulfilling prophecy. For our purposes in understanding AI, let us address the following question.

What Is Moore's law?

As it applies to AI, we will define Moore's law as follows: The data density of an integrated circuit and the associated computer performance will cost-effectively double every eighteen months. If we consider eighteen months to represent a technology generation, this means every eighteen months we receive double the data density and associated computer performance at approximately the same cost as the previous generation. Most experts, including Moore, expect Moore's law to hold for at least another two decades, but this is debatable, as I discuss later in the chapter. Below is a graphical depiction (courtesy of Wikimedia Commons) of Moore's law, illustrating transistor counts for integrated circuits plotted against their dates of introduction (1971–2011).

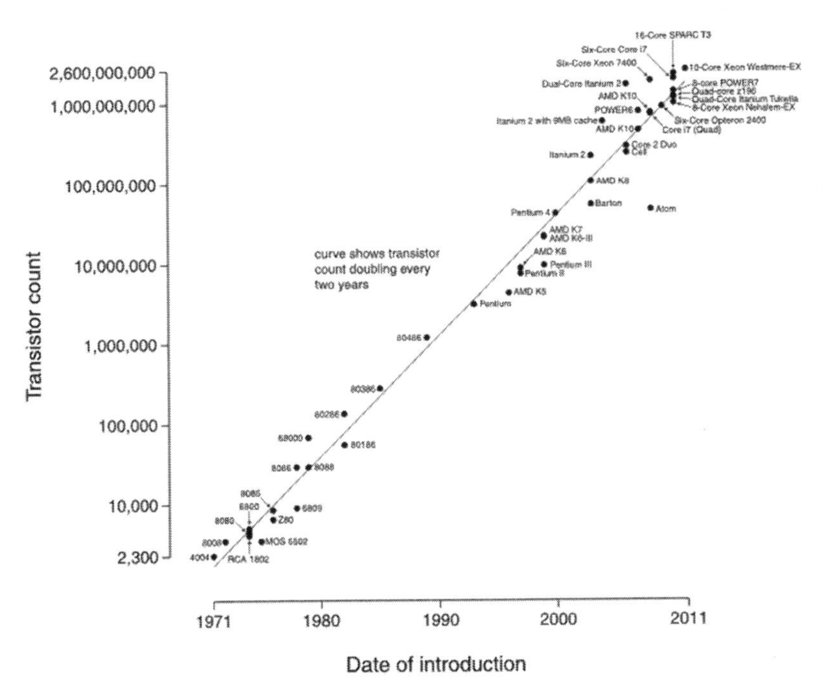

Moore's Law Illustrating Transistor Counts for Integrated Circuits

As previously mentioned, Moore's law is not a physical law of science. Rather it may be considered a trend or a general rule. This begs the following question.

How Long Will Moore's Law Hold?

There are numerous estimates regarding how long Moore's law will hold. Since it is not a physical law, its applicability is routinely questioned. For approximately the last half century, each estimate, at various points in time, has predicted that Moore's law would hold for another decade. This has been occurring for almost five decades.

In 2005 Gordon Moore stated in an interview that Moore's law "can't continue forever. The nature of exponentials is that you push them out and eventually disaster happens." Moore noted that transistors eventually would reach the limits of miniaturization at atomic levels. "In terms of size [of transistors] you can see that we're approaching the size of atoms, which is a fundamental barrier, but it'll be two or three generations before we get that far—but that's as far out as we've ever been able to see. We have another 10 to 20 years before we reach a fundamental limit."

However, new technologies are emerging to use molecules individually positioned, replacing transistors altogether. This means computer "switches" will not be transistors but molecules. The position of the molecules will be the new switches. This technology is predicted to emerge by 2020 (Baptiste Waldner, *Nanocomputers and Swarm Intelligence*, 2008).

Some see Moore's law extending far into the future. Lawrence Krauss and Glenn D. Starkman predicted an ultimate limit of around six hundred years (Lawrence M. Krauss, Glenn D. Starkman, "Universal Limits of Computation," arXiv:astro-ph/0404510, May 10, 2004).

I worked in the semiconductor industry for more than thirty years, during which time Moore's law always appeared as if it would reach an impenetrable barrier. This, however, did not happen. New technologies constantly seemed to provide a stay of execution. We know that at some point the trend may change, but no one really has made a definitive case as to when this trend will end. The difficulty in predicting the end has to do with how one interprets Moore's law. If one takes Moore's original interpretation, which defined the trend in terms of the number of transistors that could be put on an integrated circuit, the end point may be somewhere around 2018 to 2020. Defining it in terms of "data density of an integrated circuit," however, as we did regarding AI, removes the constraint of transistors and opens up a new array of technologies, including molecular positioning.

Will Moore's law hold for another decade or another six hundred years? No one really knows the answer. Most people believe that eventually the trend will end, but when and why remain unanswered questions. If it does end, and Moore's law no longer applies, another question emerges.

What Will Replace Moore's Law?

Ray Kurzweil views Moore's law in much the same way we defined it, not tied to specific technologies but rather as a "paradigm to forecast accelerating price-performance ratios." From Kurzweil's viewpoint:

> Moore's law of Integrated Circuits was not the first, but the fifth paradigm to forecast accelerating price-performance ratios. Computing devices have been consistently multiplying in power (per unit of time) from the mechanical calculating devices used in the 1890 U.S. Census, to [Newman's] relay-based "[Heath] Robinson" machine that cracked the Lorenz cipher, to the CBS vacuum tube computer that predicted the election of Eisenhower, to the transistor-based machines used

> in the first space launches, to the integrated circuit-based personal computer. (Raymond Kurzweil, "The Law of Accelerating Returns," www.KurzweilAI.net)

In the wider sense, Moore's law is not about transistors or specific technologies. In my opinion it is a paradigm related to humankind's creativity. The new computers following Moore's law may be based on some new type of technology (e.g., optical computers, quantum computers, DNA computing) that bears little to no resemblance to current integrated-circuit technology. It appears that what Moore really uncovered was humankind's ability to cost-effectively accelerate technology performance.

The next chapter explains how Moore's law gave rise to intelligent agents.

Conclusions from Chapter 3

- Moore's law is not, in the true sense, a physical law of science. It is an observation of a trend related to computer technology.
- Moore's law states that the data density of an integrated circuit and the associated computer performance will cost-effectively double every eighteen months.
- Moore's law accurately has predicted the trend in cost-effective computer technology advancements for more than half a century.
- No one really knows how long Moore's law will continue to hold. Predictions vary from decades to six hundred years.
- Moore's law may be viewed as a paradigm related to humankind's technological creativity, forecasting accelerating price-performance ratios of computer technology. From this viewpoint we can envision that future computers following Moore's law may be based on some new type of technology (e.g., optical, quantum computers, DNA computing) that bears little to no resemblance to current integrated-circuit technology.

The main lesson of thirty-five years of AI research is that the hard problems are easy and the easy problems are hard. The mental abilities of a four-year-old that we take for granted—recognizing a face, lifting a pencil, walking across a room, answering a question—in fact solve some of the hardest engineering problems ever conceived.... As the new generation of intelligent devices appears, it will be the stock analysts and petrochemical engineers and parole board members who are in danger of being replaced by machines. The gardeners, receptionists, and cooks are secure in their jobs for decades to come.

—Steven Pinker,
***The Language Instinct* (2012)**

CHAPTER 4

The Rise of Intelligent Agents

The road to intelligent machines has been difficult, filled with hairpin curves, steep hills, crevices, potholes, intersections, stop signs, and occasionally smooth and straight sections. The initial over-the-top optimism of AI founders John McCarthy, Marvin Minsky, Allen Newell, and Herbert Simon set unrealistic expectations. According to their predictions, by now every household should have its own humanoid robot to cook, clean, and do yard work and every other conceivable household task we humans perform.

During the course of my career, I have managed hundreds of scientists and engineers. In my experience they are, for the most part, overly optimistic as a group. When they say something was finished, it usually means it's in the final stages of testing or inspection. When they say they will have a problem solved in a week, it usually means a month or more. Whatever schedules they give us—the management—we normally have to pad, sometimes doubling them, before we use the schedules to plan or before we give them to our clients. It is just part of their nature to be optimistic, believing the tasks associated with the goals will go without a hitch, or the solution to a problem will be just one experiment away. Often if you ask a simple question, you'll receive the "theory of everything" as a reply. If the question relates to a problem, the answer will involve the history of humankind and fingers will be pointed in every direction. I am exaggerating slightly to make a point, but as humorous as this may sound, there is more than a kernel of truth in what I've stated.

This type of optimism accompanied the founding of AI. The founders dreamed with sugarplums in their heads, and we wanted to believe it. We wanted the world to be easier. We wanted intelligent machines to do the heavy lifting and drudgery of everyday chores. We did not have to envision it. The science-fiction writers of television series such as *Star Trek* envisioned it for us, and we wanted to believe that artificial life-forms, such as Lieutenant Commander Data on *Star Trek: The Next Generation*, were just a decade away. However, that is not what happened. The field of AI did not change the world overnight or even in a decade. Much like a ninja, it slowly and invisibly crept into our lives over the last half century, disguised behind "smart" applications.

After several starts and stops and two AI winters, AI researchers and engineers started to get it right. Instead of building a do-it-all intelligent machine, they focused on solving specific applications. To address the applications, researchers pursued various approaches for specific intelligent systems. After accomplishing that, they began to integrate the approaches, which brought us closer to artificial "general" intelligence, equal to human intelligence.

Many people not engaged in professional scientific research believe that scientists and engineers follow a strict orderly process, sometimes referred to as the "scientific method," to develop and apply new technology. Let me dispel that paradigm. It is simply not true. In many cases a scientific field is approached via many different angles, and the approaches depend on the experience and paradigms of those involved. This is especially true in regard to AI research, as will soon become apparent.

The most important concept to understand is that no unifying theory guides AI research. Researchers disagree among themselves, and we have more questions than answers. Here are two major questions that still haunt AI research.

1. Should AI simulate human intelligence, incorporating the sciences of psychology and neurology, or is human biology irrelevant?

2. Can AI, simulating a human mind, be developed using simple principles, such as logic and mechanical reasoning, or does it require solving a large number of completely unrelated problems?

Why do the above questions still haunt AI? Let us take some examples.

- Similar types of questions arose in other scientific fields. For example, in the early stages of aeronautics, engineers questioned whether flying machines should incorporate bird biology. Eventually bird biology proved to be a dead end and irrelevant to aeronautics.

- When it comes to solving problems, humans rely heavily on experience, and we augment it with reasoning. In business, for example, for every problem encountered, there are numerous solutions. The solution chosen is biased by the paradigms of those involved. If, for example, the problem is related to increasing the production of a product being manufactured, some managers may add more people to the work force, some may work at improving efficiency, and some may do both. I have long held the belief that for every problem we face in industry, there are at least ten solutions, and eight of them, although different, yield equivalent results. However, if you look at the previous example, you may be tempted to believe improving efficiency is a superior (i.e., more elegant) solution as opposed to increasing the work force. Improving efficiency, however, costs time and money. In many cases it is more expedient to increase the work force. My point is that humans approach solving a problem by using their accumulated life experiences, which may not even relate directly to the specific problem, and augment their life experiences with reasoning. Given the way human minds work, it is only natural to ask whether intelligent machines will have to approach problem solving in a similar way, namely by solving numerous unrelated problems as a path to the specific solution required.

Scientific work in AI dates back to the 1940s, long before the AI field had an official name. Early research in the 1940s and 1950s focused on attempting to simulate the human brain by using rudimentary cybernetics (i.e., control systems). Control systems use a two-step approach to controlling their environment.

1. An action by the system generates some change in its environment.

2. The system senses that change (i.e., feedback), which triggers the system to change in response.

A simple example of this type of control system is a thermostat. If you set it for a specific temperature, for example 72 degrees Fahrenheit, and the temperature drops below the set point, the thermostat will turn on the furnace. If the temperature increases above the set point, the thermostat will turn off the furnace. However, during the 1940s and 1950s, the entire area of brain simulation and cybernetics was a concept ahead of its time. While elements of these fields would survive, the approach of brain simulation and cybernetics was largely abandoned as access to computers became available in the mid-1950s.

With access to electronic digital programmable computers in the mid-1950s, AI researchers began to focus on symbol manipulation (i.e., the manipulation of mathematical expressions, as is found in algebra) to emulate human intelligence. Three institutions led the charge: Carnegie Mellon University, Stanford University, and the Massachusetts Institute of Technology (MIT). Each university had its own style of research, which the American philosopher John Haugeland (1945–2010) named "good old-fashioned AI" or "GOFAI."

From the 1960s through the 1970s, symbolic approaches achieved success at simulating high-level thinking in specific application programs. For example, in 1963, Danny Bobrow's technical report from MIT's AI group proved that computers could understand natural language well enough to solve algebra word problems correctly. The success of symbolic approaches added credence to the belief that symbolic approaches eventually would succeed in creating a machine with artificial general intelligence, also known as "strong AI," equivalent to a human mind's intelligence.

By the 1980s, however, symbolic approaches had run their course and fallen short of the goal of artificial general intelligence. Many AI researchers felt symbolic approaches never would emulate the processes of human cognition, such as perception, learning, and pattern recognition. The next step was a small retreat, and a new era of AI research termed "subsymbolic" emerged. Instead of attempting general AI, researchers turned their attention to solving smaller specific problems. For example researchers such as Australian computer scientist and former MIT Panasonic Professor of Robotics Rodney Brooks rejected symbolic AI. Instead he focused on solving engineering problems related to enabling robots to move.

In the 1990s, concurrent with subsymbolic approaches, AI researchers began to incorporate statistical approaches, again addressing specific problems. Statistical methodologies involve advanced mathematics and are truly scientific in that they are both measurable and verifiable. Statistical approaches proved to be a highly successful AI methodology. The advanced mathematics that underpin statistical AI enabled collaboration with more established fields, including mathematics, economics, and operations research. Computer scientists Stuart Russell and Peter Norvig describe this movement as the victory of the "neats" over the "scruffies," two major opposing schools of AI research. Neats assert that AI solutions should be elegant, clear, and provable. Scruffies, on the other hand, assert that intelligence is too complicated to adhere to neat methodology.

From the 1990s to the present, despite the arguments between neats, scruffies, and other AI schools, some of AI's greatest successes have been the result of combining approaches, which has resulted in what is known as the "intelligent agent." The intelligent agent is a system that interacts with its environment and takes calculated actions (i.e., based on their success probability) to achieve its goal. The intelligent agent can be a simple system, such as a thermostat, or a complex system, similar conceptually to a human being. Intelligent agents also can be combined to form multiagent systems, similar conceptually to a large corporation, with a hierarchical control system to bridge lower-level subsymbolic AI systems to higher-level symbolic AI systems.

The intelligent-agent approach, including integration of intelligent agents to form a hierarchy of multiagents, places no restriction on the AI methodology employed to achieve the goal. Rather than arguing philosophy, the emphasis is on achieving results. The key to achieving the greatest results has proven to be integrating approaches, much like a symphonic orchestra integrates a variety of instruments to perform a symphony.

In the last seventy years, the approach to achieving AI has been more like that of a machine gun firing broadly in the direction of the target than a well-aimed rifle shot. In fits of starts and stops, numerous schools of AI research have pushed the technology forward. Starting with the loftiest goals of emulating a human mind, retreating to solving specific well-defined problems, and now again aiming toward artificial general intelligence, AI research is a near-perfect example of all human technology development, exemplifying trial-and-error learning, interrupted with spurts of genius.

Although AI has come a long way in the last seventy years and has been able to equal and exceed human intelligence in specific areas, such as playing chess, it still falls short of general human intelligence or strong AI. There are two significant problems associated with strong AI. First, we need a machine with processing power equal to that of a human brain. Second, we need programs that allow such a machine to emulate a human brain. In the next chapter, I address the first challenge, namely developing a machine with raw processing power equal to that of a human brain.

Conclusions from Chapter 4

- No unifying theory guides artificial intelligence (AI) research.

- AI research dates back to the 1940s and early 1950s and focused on attempting to simulate the human brain, using rudimentary cybernetics (i.e., control systems).

- In the mid-1950s, AI researchers, using computers, began to focus on symbol manipulation (i.e., the manipulation of mathematical expressions, such as in algebra) to emulate human intelligence.

- From the 1960s through the 1970s, symbolic approaches achieved success at simulating high-level thinking in specific applications, fueling the belief that symbolic approaches eventually would lead to artificial general intelligence or strong AI, emulating human intelligence.

- By the 1980s symbolic approaches had run their course, falling short of the goal of artificial general intelligence. The next step was a small retreat, and a new era of AI research termed "subsymbolic" emerged. Instead of attempting general AI, researchers turned their attention to solving smaller specific problems.

- In the 1990s, concurrent with subsymbolic approaches, AI researchers began to incorporate statistical approaches, again addressing specific problems and enabling collaboration with more established fields, including mathematics, economics, and operations research.

- From the 1990s to the present, AI's greatest successes have been the result of combining approaches, which has resulted in what is known as the "intelligent agent." The intelligent agent is a system that interacts with its environment and takes calculated actions to achieve its goal. Intelligent agents also can be combined to form multiagent systems, with a hierarchical control system to bridge lower-level subsymbolic AI systems to higher-level symbolic AI systems.

- Although AI has come a long way in the last seventy years and has been able to equal and exceed human intelligence in specific areas, such as playing chess, it still falls short of general human intelligence or strong AI (i.e., being completely equal to the human mind).

The human brain has about 100 billion neurons. With an estimated average of one thousand connections between each neuron and its neighbors, we have about 100 trillion connections, each capable of a simultaneous calculation... [but] only 200 calculations per second... With 100 trillion connections, each computing at 200 calculations per second, we get 20 million billion calculations per second. This is a conservatively high estimate.... By the year 2020, [a massively parallel neural net computer] will have doubled about 23 times (from 1997's $2,000 modestly parallel computer that could perform around 2 billion connection calculations per second)... resulting in a speed of about 20 million billion neural connection calculations per second, which is equal to the human brain.

—Ray Kurzweil,
***The Age of Spiritual Machines* (1999)**

CHAPTER 5

Raw Processing Power Equals a Human Brain

If we want to view the human brain in terms of a computer, one approach would be to take the number of calculations per second that an average human brain is able to process and compare that with today's best computers. This is not an exact science. No one really knows how many calculations per second an average human brain is able to process, but some estimates (www.rawstory.com/rs/2012/06/18/earths-supercomputing-power-surpasses-human-brain-three-times-over) suggest it is in the order of 36.8 petaflops of data (a petaflop is equal to one quadrillion calculations per second). Let us compare the human brain's processing power with the best current computers on record, listed below by year and processing-power achievement.

- **June 18, 2012:** IBM's Sequoia supercomputer system, based at the US Lawrence Livermore National Laboratory (LLNL), reached sixteen petaflops, setting the world record and claiming first place in the latest TOP500 list (a list of the top five hundred computers ranked by a benchmark known as LINPACK (related to their ability to solve a set of linear equations) to decide whether they qualify for the TOP500.

- **November 12, 2012:** The TOP500 list certified Titan as the world's fastest supercomputer per the LINPACK benchmark, at 17.59 petaflops. Cray Incorporated, at the Oak Ridge National Laboratory, developed it.

- **June 10, 2013:** China's Tianhe-2 was ranked the world's fastest supercomputer, with a record of 33.86 petaflops.

Using Moore's law, discussed in chapter 3, we can extrapolate that in terms of raw processing power (petaflops), computer processing power will meet or exceed that of the human mind by about 2015 to 2017. This does not mean that by 2017 we will have a computer that is equal to the human mind. Software plays a key role in both processing power (MIPS) and AI.

To understand the critical role that software plays, we must understand what we are asking AI to accomplish in emulating human intelligence. Here is a thumbnail sketch of the capabilities that researchers consider necessary.

- **Reasoning:** step-by-step reasoning that humans use to solve problems or make logical decisions
- **Knowledge:** extensive knowledge, similar to what an educated human would possess
- **Planning:** the ability to set goals and achieve them
- **Learning:** the ability to acquire knowledge through experience and use that knowledge to improve
- **Language:** the ability to understand the languages humans speak and write
- **Moving:** the ability to move and navigate, including knowing where it is relative to other objects and obstacles
- **Manipulation:** the ability to secure and handle an object
- **Vision:** the ability to analyze visual input, including facial and object recognition
- **Social intelligence:** the ability to recognize, interpret, and process human psychology and emotions and respond appropriately

- **Creativity:** the ability to generate outputs that can be considered creative or the ability to identify and assess creativity

This list makes clear that raw computer processing and sensing are only two elements in emulating the human mind. Obviously software is also a critical element. Each of the capabilities delineated above requires a computer program. To emulate a human mind, the computer programs would need to act both independently and interactively, depending on the specific circumstance.

In terms of raw computer processing, with the development of China's Tianhe-2 computer, we are on the threshold of having a computer with the raw processing power of a human mind. The development of a computer that will emulate a human mind, however, may still be one, two, or even more decades away, due to software and sensing requirements.

How close are we to creating a computer with strong AI that emulates the human mind? The answer is that no one knows. At any given point in time, over the last fifty years of AI research, the goal of developing a computer with strong AI that would emulate a human mind has appeared close, perhaps as little as a decade away. Even with the relentless progress in computer technology, however, this goal continues to remain elusive.

How will we know when we actually have reached the goal of creating a computer with strong AI that emulates the human mind? In 1950 Alan Turing proposed the Turing test as a methodology to test the intelligence of an agent. The Turing test requires that a human "judge" engage both a human and a computer with strong AI in a natural-language conversation. None of the participants, however, can see each other. If the judge is unable to distinguish between the human and a computer with strong AI, the computer with strong AI will be said to have passed the Turing test. This test does not require that the answers be correct, just indistinguishable. Passing the Turing test requires almost all the major capabilities associated with strong AI to be equivalent to those of a human brain. It is an extremely difficult test, and to date no intelligent agent has passed it.

From the late 1990s to the present, new tests have been developed to specifically evaluate the intelligence of machines. These new tests have been based on the mathematical definitions of intelligence, and they have eliminated human testers. These tests have claimed to be "beyond" or superior to the Turing test. It is too early, however, to evaluate their acceptance within the scientific community. I mention them for completeness. The Turing test still stands as the gold standard that these and other tests use as a benchmark.

Today no intelligent agent or system of intelligent agents is able to pass the Turing test. Comparisons with human performance, however, are being evaluated on constrained and well-defined problems, known as "expert" Turing tests, which were first proposed by computer scientist Edward Feigenbaum in a 2003 *Journal of the ACM* paper titled, "Some Challenges and Grand Challenges for Computational Intelligence." The results of these tests are categorized below, along with some application examples in which computer AI is able to meet the challenges in the category.

- **Optimal:** It is not possible to perform better. Computer AI examples include tic-tac-toe, which can be played to a draw by any player, and the Rubik's Cube, which is solvable by following a specific algorithm.

- **Strong-superhuman:** These computers perform better than all humans. Computer AI examples include Scrabble and quiz-show question answering. Scrabble belongs to the category of "solved games" that can be correctly predicted from any position, given that both players play perfectly. Scrabble was solved by Alan Frank using the *Official Scrabble Players Dictionary* in

1987. Quiz-show question answering relies on the enormous databases stored on computer hard drives combined with the computer's mechanical reasoning to reach the correct answer. A good example of this appears in chapter 2 of this book. ("In 2011 on an exhibition match on the popular TV quiz show *Jeopardy!*, an IBM computer named Watson defeated *Jeopardy!*'s greatest champions, Brad Rutter and Ken Jennings.")

- **Superhuman:** These computers perform better than most humans. Computer AI examples include chess and crosswords. Chess belongs to the category of "perfect-play games" (i.e., the behavior or strategy of a player is what leads to the best possible outcome for that player regardless of the response by the opponent). Chess arguably overlaps into the strong-superhuman category because a chess-playing computer is able to benefit from certain endgame positions (in the form of endgame table bases), which allow it to play perfectly from a specific point in the endgame. Backgammon falls into the category of "unsolved games," having 10^{18} (i.e., a "1" with eighteen zeros after it) positions, which makes it difficult to solve via brute-force searching, similar to what was done for the endgame positions in chess. The best backgammon programs, however, rank among the top twenty players in the world.

- **Par-human:** These machines perform similarly to most humans. Computer AI examples include standardized (ISO 1073-1:1976) optical character recognition and Go. The font used for optical character recognition was optimized using simple, thick strokes to form recognizable characters that humans and computers could easily recognize. Go is a two-player board game that originated in China more than 2,500 years ago and belongs to the category of "partially solved games," for which only some positions have been solved. Go has deceptively simple rules but is rich in strategy, prompting chess master Emanuel Lasker to state, "The rules of Go are so elegant, organic, and rigorously logical that if intelligent life-forms exist elsewhere in the universe, they almost certainly play Go."

- **Subhuman:** These computers perform worse than most humans. Computer AI examples include handwriting recognition and translation. Handwriting differs from human to human, and there are numerous cases in which humans cannot read their own handwriting. Handwriting recognition entails optical character recognition but without standardization, making the task especially difficult. In addition a complete handwriting recognition system also must handle formatting, perform correct segmentation into characters, and find the most plausible words. Humans outperform computers in this area. Translation is difficult, even for humans. Not all languages have the same content. There are words in some languages that have no equivalent word in another. In addition, while computers are able to mechanically translate from one language to another, resulting in a "rough draft," the sense, meaning, and impact may be lost. Factually, communication in human language is context embedded, requiring a person to comprehend the context. For these reasons, humans are typically better translators than computers.

Given the above discussion, let us return to our original question: What is required for a computer to emulate a human brain's intelligence? From a strictly hardware viewpoint, many experts predict that within a decade computers will match the processing power of a human mind. This does not mean, however, that they will emulate a human brain's intelligence.

What is required for a computer to emulate a human brain's intelligence? To answer that question, we must first answer a more fundamental question, namely "What is human intelligence?" There is no one widely accepted answer. Here are two definitions that have found some acceptance among the scientific community.

1. "A very general mental capability that, among other things, involves the ability to reason, plan, solve problems, think abstractly, comprehend complex ideas, learn quickly, and learn from experience. It is not merely book learning, a narrow academic skill, or test-taking smarts. Rather, it reflects a broader and deeper capability for comprehending our surroundings—'catching on,' 'making sense' of things, or 'figuring out' what to do" ("Mainstream Science on Intelligence," an editorial statement by fifty-two researchers, *The Wall Street Journal,* December 13, 1994).

2. "Individuals differ from one another in their ability to understand complex ideas, to adapt effectively to the environment, to learn from experience, to engage in various forms of reasoning, to overcome obstacles by thinking. Although these individual differences can be substantial, they are never entirely consistent: a given person's intellectual performance will vary on different occasions, in different domains, as judged by different criteria. Concepts of 'intelligence' are attempts to clarify and organize this complex set of phenomena. Although considerable clarity has been achieved in some areas, no such conceptualization has yet answered all the important questions, and none commands universal assent. Indeed, when two dozen prominent theorists were recently asked to define intelligence, they gave two dozen, somewhat different, definitions."

("Intelligence: Knowns and Unknowns," a report published by the Board of Scientific Affairs of the American Psychological Association, 1995).

These two definitions provide some indication of the difficulty encountered in attempting to define human intelligence. However, both definitions, among other traits, require the ability to learn from experience. Understanding complex ideas, solving problems, and adapting to the environment are insufficient to define human intelligence. Up to this point in the book, I have shown that intelligent machines can grasp problems (via symbolic and subsymbolic approaches), solve problems, and adapt to their environment. I have not, however, delineated how intelligent machines will learn from experience. This is the topic of the next chapter.

Conclusions from Chapter 5

- There is no one widely accepted answer as to what constitutes human intelligence, which makes defining when an intelligent machine will emulate a human brain difficult.

- In 1950 Alan Turing proposed the Turing test as a methodology to test the intelligence of an agent. The Turing test requires that a human judge engage a human as well as a computer with strong AI in a natural-language conversation. None of the participants, however, can see each other. If the judge is unable to distinguish between the human and the computer with strong AI, the computer with strong AI will be said to have passed the Turing test.

- Today no intelligent agent or system of intelligent agents is able to pass the Turing test.

- Comparisons with human performance, however, are being evaluated on constrained and well-defined problems. These are known as expert Turing tests.

    - **Optimal:** It is not possible to perform better than these computers.

    - **Strong-superhuman:** These computers perform better than all humans.

    - **Superhuman:** These perform better than most humans.

   - **Par-human:** These perform similarly to most humans.

   - **Subhuman:** These perform worse than most humans.

- From a strictly hardware viewpoint, we are on the threshold of developing a computer that matches the processing power of the human mind.

Field of study that gives computers the ability to learn without being explicitly programmed.

—Arthur Samuel's definition of machine learning (1959)

CHAPTER 6

Self-Learning Machines

How is it possible to wire together microprocessors, hard drives, memory chips, and numerous other electronic hardware components and create a machine that will teach itself to learn?

Let us start by defining machine learning. The most widely accepted definition comes from Tom M. Mitchell, a American computer scientist and E. Fredkin University Professor at Carnegie Mellon University. Here is his formal definition: "A computer program is said to learn from experience E with respect to some class of tasks T and performance measure P, if its performance at tasks in T, as measured by P, improves with experience E." In simple terms machine learning requires a machine to learn similar to the way humans do, namely from experience, and continue to improve its performance as it gains more experience.

Machine learning is a branch of AI; it utilizes algorithms that improve automatically through experience. Machine learning also has been a focus of AI research since the field's inception. There are numerous computer software programs, known as machine-learning algorithms, that use various computational techniques to predict outcomes of new, unseen experiences. The algorithms' performance is a branch of theoretical computer science known as "computational learning theory." What this means in simple terms is that an intelligent machine has in its memory data that relates to a finite set of experiences. The machine-learning algorithms (i.e., software) access this data for its similarity to a new experience and use a specific algorithm (or combination of algorithms) to guide the machine to predict an outcome of this new experience. Since the experience data in the machine's memory is limited, the algorithms are unable to predict outcomes with certainty. Instead they associate a probability to a specific outcome and act in accordance with the highest probability. Optical character recognition is an example of machine learning. In this case the computer recognizes printed characters based on previous examples. As anyone who has ever used an optical character-recognition program knows, however, the programs are far from 100 percent accurate. In my experience the best case is a little more than 95 percent accurate when the text is clear and uses a common font.

There are eleven major machine-learning algorithms and numerous variations of these algorithms. To study and understand each would be a formidable task. Fortunately, though, machine-learning algorithms fall into three major classifications. By understanding these classifications, along with representative examples of algorithms, we can gain significant insight into the science of machine learning. Therefore let us review the three major classifications and some representative examples.

1. **Supervised learning:** This class of algorithms infers a function (a way of mapping or relating an input to an output) from training data, which consists of training examples. Each example consists of an input object and a desired output value. Ideally the inferred function (generalized from the training data) allows the algorithm to analyze new data (unseen instances/inputs) and map it to (i.e., predict) a high-probability output.

   Examples of supervised learning algorithms: Numerous algorithms are used in supervised leaning. Two examples are decision trees and neural networks.

   - A decision tree is a treelike graph that models decisions and their possible consequences. It is displayable as a graphical depiction of a computer algorithm (i.e., a computer program), and it is sometimes used to guide the development of the algorithm. Decision-tree learning is a method that uses a decision tree as a predictive model. It uses observations about an item to select a specific path through the decision tree. Using this technique, a computer using decision-tree learning can reach conclusions. The conclusions it reaches have a specific probability of being correct. Decision-tree learning finds application in three fields: statistics, data mining, and machine learning. Figure 1 is an example of a simple decision tree that attempts to answer whether an animal is a dog or a cat.

   - This is an extremely simple example, and there are numerous ways to draw a decision tree. Notice that the conclusions are probabilities, indicated by the word "probably." In a computer program, the conclusions would have numerical

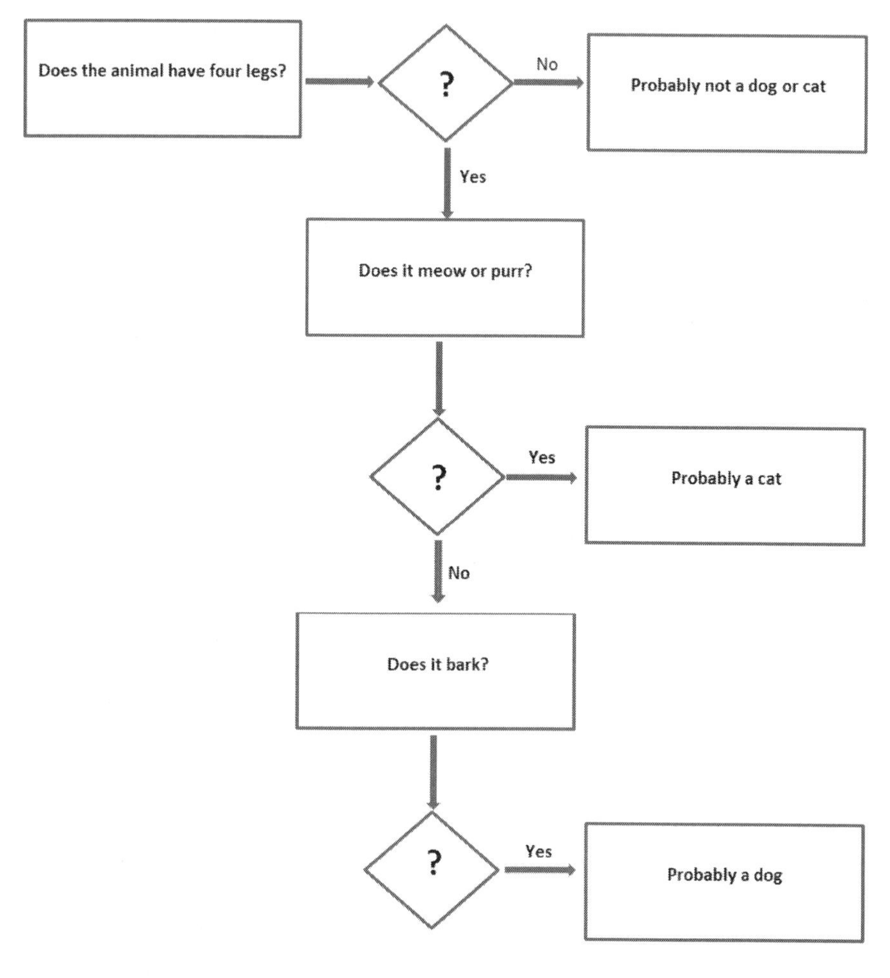

Simple Decision Tree Flow Chart Example

values associated with their probability. This simple decision tree has limited capabilities to reach conclusions. For example it is possible to have a three-legged dog or cat; in this case the decision tree would not reach a correct conclusion. A sophisticated computer program uses feedback to improve the algorithm's predictive capability, which is called "machine learning."

- Neural networks are used in both supervised and unsupervised learning. A neural network attempts to model the biological functioning of the human brain. For example, using the human brain as a model, the units in the network (i.e., artificial neurons) could represent human-brain neurons, and the connections between the artificial neurons could represent human-brain synapses. In this model the interconnected artificial neurons process information to perform a computation. In machine learning the neural network typically is adaptive and able to change its structure during the learning phase, which makes it a self-learning computer capable of modeling complex relationships to find patterns in data (for unsupervised learning) and predict a likely outcome (for supervised learning).

2. **Unsupervised learning:** This class of algorithms seeks to find hidden structures (patterns in data) in a stream of input (unlabeled data). Unlike in supervised learning, the examples presented to the learner are unlabeled, which makes it impossible to assign an error or reward to a potential solution. Numerous algorithms are used in unsupervised learning, including neural networks, as discussed above. Another typical algorithm used in unsupervised learning is association-rule learning, which is described below.

   Example of an unsupervised learning algorithm: Association-rule learning is a method for discerning relationships between variables in databases (information depositories). Association rules are if/then statements—an antecedent (if) and a consequent (then)—that uncover relationships between seemingly unrelated data in a database. Here is an example of an association rule: "If customers buy bacon, they are 80 percent likely to buy eggs." Association rules are developed by analyzing data for frequent if/then patterns, along with "support" (the frequency that the items appear in the database) and "confidence" (the number of times the if/then statements are true) criteria. Programmers use association rules to build programs capable of machine learning. There are numerous algorithms for generating association rules, but as far as machine learning goes, they perform in a similar fashion. They analyze the data for frequent if/then patterns, apply "support" and "confidence" criteria, and use feedback to update the criteria. The goal in machine learning is to discover interesting (highly probable and other differentiating measures) relationships between variables in databases. These relationships allow an intelligent machine presented with an "if" antecedent to predict a "then" consequence with a probability level. The "beer and diaper" anecdotal story is often cited as an example of an association rule to illustrate how two seemingly unrelated items are associated. It goes something like this: Between 5:00 and 7:00 p.m., customers (presumably men) who buy diapers tend also to buy beer (Daniel Powers, www.dssresources.com/newsletters/66.php).

3. **Reinforcement learning:** Reinforcement learning was inspired by behavioral psychology. It focuses on which actions an agent (an intelligent machine) should take to maximize a reward (for example a numerical value associated with utility). In effect the agent receives rewards for good responses and punishment for bad ones. The algorithms for reinforcement learning require the agent to take discrete time steps and calculate the reward as a function of having taken that step. At this point the agent takes another time step and again calculates the reward, which provides feedback to guide the agent's

next action. The agent's goal is to collect as much reward as possible.

Example of a reinforcement-learning algorithm: Although there are numerous reinforcement-learning algorithms, one of the most popular is temporal-difference (TD) learning, a predictive method. Essentially the agent makes predictions about an outcome but assumes preceding predictions related to that outcome are correlated. Weather forecasting is a simple example of reinforcement learning. Assume it is Monday and you'd like to predict what the weather will be on Saturday. Using temporal-difference learning, you would predict Saturday's weather each day up to and including Friday. However, you would use the data gathered daily to update your prediction for Saturday's weather. For example each day you may track cold and warm fronts moving in and out, along with relative humidity. By Friday the previous predictions will provide a basis to predict Saturday's weather).

In essence machine learning incorporates four essential elements.

1. **Representation:** The intelligent machine must be able to assimilate data (input) and transform it in a way that makes it useful for a specific algorithm.

2. **Generalization:** The intelligent machine must be able to accurately map unseen data to similar data in the learning data set.

3. **Algorithm selection:** After generalization the intelligent machine must choose and/or combine algorithms to make a computation (such as a decision or an evaluation).

4. **Feedback:** After a computation, the intelligent machine must use feedback (such as a reward or punishment) to improve its ability to perform steps 1 through 3 above.

Machine learning is similar to human learning in many respects. The most difficult issue in machine learning is generalization or what is often referred to as abstraction. This is simply the ability to determine the features and structures of an object (i.e., data) relevant to solving the problem. Humans are excellent when it comes to abstracting the essence of an object. For example, regardless of the breed or type of dog, whether we see a small, large, multicolor, long-hair, short-hair, large-nose, or short-nose animal, we immediately recognize that the animal is a dog. Most four-year-old children immediately recognize dogs. However, most intelligent agents have a difficult time with generalization and require sophisticated computer programs to enable them to generalize.

Machine learning has come a long way since the 1972 introduction of Pong, the first game developed by Atari Inc. Today's computer games are incredibly realistic, and the graphics are similar to watching a movie. Few of us can win a chess game on our computer or smartphone unless we set the difficulty level to low. In general machine learning appears to be accelerating, even faster than the field of AI as a whole. We may, however, see a bootstrap effect, in which machine learning results in highly intelligent agents that accelerate the development of artificial general intelligence, but there is more to the human mind than intelligence. One of the most important characteristics of our humanity is our ability to feel human emotions.

This raises an important question. When will computers be capable of feeling human emotions? A new science is emerging to address how to develop and program computers to be capable of simulating and eventually feeling human emotions. This new science is termed "affective computing." Do you want to know more?

Conclusions from Chapter 6

- Tom M. Mitchell's formal definition of machine learning is as follows. "A computer program is said to learn from experience E with respect to some class of tasks T and performance measure P, if its performance at tasks in T, as measured by P, improves with experience E."

- Machine learning is a branch of theoretical computer science known as computational learning theory.

- Machine-learning algorithms (computer programs that enable a machine to learn) fall into three major classifications.

    1. **Supervised learning:** This class of algorithms infers a function (a way of mapping or relating an input to an output) from training data, which consists of training examples.

    2. **Unsupervised learning:** This class of algorithms seeks to find hidden structures (patterns in data) in a stream of input (unlabeled data).

    3. **Reinforcement learning:** This focuses on which actions an agent (i.e., an intelligent machine) should take to maximize a reward (for example, a numerical value associated with utility).

- In essence machine learning incorporates four essential elements.

    1. **Representation:** The intelligent machine must be able to assimilate data (input) and transform it in a way that makes it useful for a specific algorithm.

2. **Generalization:** The intelligent machine must be able to accurately map unseen data to similar data in the learning data set.

3. **Algorithm selection:** After generalization the intelligent machine must choose and/or combine algorithms to make a computation (such as a decision or an evaluation).

4. **Feedback:** After a computation the intelligent machine must use feedback (such as a reward or punishment) to improve its ability to perform steps 1 through 3 above.

- Machine learning is similar to human learning in many respects. The most difficult issue in machine learning is generalization or what is often referred to as abstraction.

- In general, machine learning appears to be accelerating, even faster than the field of AI as a whole. We may, however, see a bootstrap effect, in which machine learning results in highly intelligent agents that accelerate the development of artificial general intelligence.

The office of drama is to exercise, possibly to exhaust, human emotions. The purpose of comedy is to tickle those emotions into an expression of light relief; of tragedy, to wound them and bring the relief of tears. Disgust and terror are the other points of the compass.

—Laurence Olivier (1907–1989); renowned British actor, director, and producer

CHAPTER 7

Affective Machines

Affective computing is a relatively new science. It is the science of programming computers to recognize, interpret, process, and simulate human affects. The word "affects" refers to the experience or display of feelings or emotions.

While AI has achieved superhuman status in playing chess and quiz-show games, it does not have the emotional equivalence of a four-year-old child. For example a four-year-old may love to play with toys. The child laughs with delight as the toy performs some function, such as a toy cat meowing when it is squeezed. If you take the toy away from the child, the child may become sad and cry. Computers are unable to achieve any emotional response similar to that of a four-year-old child. Computers do not exhibit joy or sadness. Some researchers believe this is actually a good thing. The intelligent machine processes and acts on information without coloring it with emotions. When you go to an ATM, you will not have to argue with the ATM regarding whether you can afford to make a withdrawal, and a robotic assistant will not lose its temper if you do not thank it after it performs a service. Highly meaningful human interactions with intelligent machines, however, will require that machines simulate human affects, such as empathy. In fact some researchers argue that machines should be able to interpret the emotional state of humans and adapt their behavior accordingly, giving appropriate responses for those emotions. For example if you are in a state of panic because your spouse is apparently having a heart attack, when you ask the machine to call for medical assistance, it should understand the urgency. In addition it will be impossible for an intelligent machine to be truly equal to a human brain without the machine possessing human affects. For example how could an artificial human brain write a romance novel without understanding love, hate, and jealousy?

Progress in regard to the development of computers with human affects has been slow. In fact this particular computer science originated with Rosalind Picard's 1995 paper on affective computing ("Affective Computing," MIT Technical Report #321, abstract, 1995). The single greatest problem involved in developing and programming computers to emulate the emotions of the human brain is that we do not fully understand how emotions are processed in the human brain. We are unable to pinpoint a specific area of the brain and scientifically argue that it is responsible for specific human emotions, which has raised questions. Are human emotions byproducts of human intelligence? Are they the result of distributed functions within the human brain? Are they learned, or are we born with them? There is no universal agreement regarding the answers to these questions. Nonetheless work on studying human affects and developing affective computing is continuing.

There are two major focuses in affective computing.

1. **Detecting and recognizing emotional information:** How do intelligent machines detect and recognize emotional information? It starts with sensors, which capture data regarding a subject's physical state or behavior. The information gathered is processed using several affective computing technologies, including speech recognition, natural-language processing, and facial-expression detection. Using sophisticated algorithms, the intelligent machine predicts the subject's affective state. For example the subject may be predicted to be angry or sad.

2. **Developing or simulating emotion in machines:** While researchers continue to develop intelligent machines with innate emotional capability, the technology is not to the level where this goal is achievable. Current technology, however, is capable of simulating emotions. For example when you provide information to a computer that is routing your telephone call, it may simulate gratitude and say, "Thank you." This has proved useful in facilitating satisfying interactivity between humans and machines. The simulation of human emotions, especially in computer-synthesized speech, is improving continually. For example you may have noticed when ordering a prescription by phone that the synthesized computer voice sounds more human as each year passes.

It is natural to ask which technologies are being employed to get intelligent machines to detect, recognize, and simulate human emotions. I will discuss them shortly, but let me alert you to one salient feature. All current technologies are based on human behavior and not on how the human mind works. The main reason for this approach is that we do not completely understand how the human mind works when it comes to human emotions. This carries an important implication. Current technology can detect, recognize, simulate, and act accordingly based on human behavior, but the machine does not feel any emotion. No matter how convincing the conversation or interaction, it is an act. The machine feels nothing. This is the result of how we have built and programmed intelligent machines, which will become evident as I discuss the seven major technologies associated with affective computing.

1. **Speech-affect detection:** The old adage "It's not what you say, but how you say it" is especially true regarding detecting and simulating human emotions. Because of this fact, it is possible to program intelligent systems to recognize changes in specific features of speech. For example our speech becomes faster, louder, and precisely enunciated with a higher and wider pitch range when we are in a state of fear, anger, or joy, and slower, lower pitched, and slurred when we are tired, bored, or sad. By analyzing speech patterns, rhythm, stress, and intonation, speech-affect detection has an average 63 percent success rate in identifying the affective state. A 63 percent success rate may appear low, but it is actually better than humans' success rate at identifying emotions. In addition numerous speech characteristics are independent of semantics or culture, which widens the application for this affect-detection technique.

2. **Facial-affect detection:** In the late 1960s, Paul Ekman, an American psychologist and pioneer in the study of facial expressions, proposed that facial expressions of emotion are not culturally determined but universal. This theory argues that human emotions have their origin in biology. In other words, we are born with them. In approximately the same

time frame, Swedish anatomist Carl-Herman Hjortsjö began to develop a facial-action coding system (FACS) that essentially defined which set of facial muscles corresponds to various human emotions. In 1972 Ekman suggested six categories of basic human emotions:

1. Anger
2. Disgust
3. Fear
4. Happiness
5. Sadness
6. Surprise

In 1978 Ekman and Wallace V. Friesen built upon Hjortsjö's FACS and developed action units (AUs) that relate to a contraction or a relaxation of one or more facial muscles associated with a human emotion. This was sufficient to form an emotional identification system that, in theory, could be used in facial-recognition programs to determine human emotions. In 1990 Ekman expanded his list of basic emotions to include:

7. Amusement
8. Contempt
9. Contentment
10. Embarrassment
11. Excitement
12. Guilt
13. Pride in achievement
14. Relief
15. Satisfaction
16. Sensory pleasure
17. Shame

Ekman initially had argued that the original six basic emotions would result in unique facial expressions. With the 1990 addition of eleven more human emotions, however, he conceded that not all emotions would be encoded in facial muscles, which provides insight into the difficulty intelligent machines would have in interpreting human emotions. In addition there were numerous other obstacles, some related to hardware and others to software. For example intelligent machines programmed to use action units work better if the facial recognition includes a full frontal view of the face. A slight twenty-degree turn of the head to the right or left, however, results in significant facial-recognition issues. Some researchers believe, though, that with the ongoing improvements in hardware and software, eventually facial-recognition software will become highly reliable. Currently, however, the use of FACS and associated action units is not reliable enough to enable widespread use in determining human emotions. Before we can consider the abilities of a computer becoming equivalent to those of a human brain, affective computing will need to be highly reliable in detecting and appropriately reacting to human emotions.

3. **Gesture-affect detection:** This is commonly referred to as "body language." We all do it without usually being aware that we are doing it. For example you may roll your eyes when you are skeptical, or you may shrug when you do not know the answer to a question. Gestures may be obvious, such as waving your hand to say hello, or subtle, such as an eye twitch when you are nervous. There are also body-language clusters, which refer to a series of gestures that a person exhibits during a specific state of mind. For example you may wave your

hand and smile to say hello to someone you recognize at a distance. Gesture-affect detection greatly increases the chances of detecting the emotional state of a subject, especially when used in conjunction with speech and face recognition.

There are two major methods being used to detect the body gestures.

- Three-dimensional model-based methods that use three-dimensional information of key elements of the body parts in order to detect body gestures

- Appearance-based methods that use images, such as hand gestures, to detect body gestures

4. **Blood volume pulse:** A subject's blood volume pulse (BVP) is a measure of blood flow through the extremities. When a subject experiences fear, his or her heart usually beats quickly. This is measured using a process called photoplethysmography, which consists of shining infrared light on the subject's skin and measuring the reflected light. The reflected light correlates to BVP. When the subject calms down, the heart rate returns to normal. This technique is not completely reliable, however, since numerous other factors can affect BVP—such as the subject being hot or cold—that have no correlation to the subject's emotional state.

5. **Facial electromyography:** This technique measures the electrical activity of the facial muscles by amplifying the electrical impulses that are generated by muscle fibers when they contract. There are two main facial muscle groups that are usually studied to detect emotion.

- The corrugator supercilii muscle, known as the "frowning" muscle, which draws the brow down into a frown, indicative of a negative, unpleasant emotional response

- The zygomaticus major muscle, responsible for pulling the corners of the mouth back when a person smiles, indicative of a positive emotional response

6. **Galvanic skin response:** This is a measure of skin conductivity, which correlates with skin moisture or perspiration. Various nervous states cause a subject to sweat, which can be correlated to galvanic skin response (GSR). This is not completely reliable, however, since other conditions, such as the subject being hot, can cause the subject to sweat and increase the GSR.

7. **Visual aesthetics:** The saying "Beauty is in the eye of the beholder" is true. This raises the following question: How can an intelligent machine appreciate visual aesthetics (i.e., beauty)? The answer is that it cannot. Once again the machine "feels" nothing. Rating visual aesthetics is accomplished by having the intelligent machine compare elements of a visual image to a peer-rated online photo-sharing website as a data source.

As mentioned earlier, machines only give the appearance of feeling emotion. In reality machines feels nothing. However, intelligent machines using simulated human affects have found numerous applications in the fields of e-learning, psychological health services, robotics, and digital pets.

It is only natural to ask, "Will an intelligent machine ever feel human affects?" This question raises a broader question: "Will an intelligent machine ever be able to completely replicate a human mind?" Experts disagree. Some experts—such as English mathematical physicist, recreational mathematician, and philosopher Roger Penrose—argue there is a limit as to what intelligent machines can do. Most experts, however, including Ray Kurzweil, argue that it will eventually be technologically feasible to copy the brain directly into an intelligent machine and that such a simulation will be identical to the original. The implication is that the intelligent machine will be a mind and be self-aware.

This begs one big question: "When will the intelligent machines become self-aware?" The time is closer than you may expect. Curious?

Conclusions from Chapter 7

- Affective computing is the science of programming computers to recognize, interpret, process, and simulate human affects (the experience or display of feelings or emotions).

- While AI has achieved superhuman status in playing chess and quiz-show games, it does not have the emotional equivalence of a four-year-old child.

- Progress in developing computers with human affects has been slow. In fact the computer science originated with Rosalind Picard's 1995 paper on affective computing.

- All current technologies to simulate human affects are based on human behavior and not on how the human mind works.

- The main reason for this approach is that we do not completely understand how the human mind works when it comes to human emotions.

- Current affects-computing technology can detect, recognize, simulate, and act accordingly based on human behavior, but the machine does not feel any emotion.

- There are seven technologies being used to simulate human affects.

    1. Speech-affect detection

    2. Facial-affect detection

    3. Gesture-affect detection

4. Blood volume pulse
5. Facial electromyography
6. Galvanic skin response
7. Visual aesthetics

- Intelligent machines that use simulated human affects have found numerous applications in the fields of e-learning, psychological health services, robotics, and digital pets.
- It is only natural to ask, "Will an intelligent machine ever feel human affects?" Experts disagree, but most experts, including Ray Kurzweil, argue that it eventually will be technologically feasible to copy the brain directly into an intelligent machine and that such a simulation will be identical to the original.
- The implication is that an intelligent machine that exactly replicates a human brain will be a mind, feel human emotions, and be conscious and even self-aware.

I think, therefore I am.

—René Descartes (1596–1650;
French philosopher, mathematician, and writer

CHAPTER 8

Self-Aware Machines

A generally accepted definition is that a person is conscious if that person is aware of his or her surroundings. If you are self-aware, it means you are self-conscious. In other words you are aware of yourself as an individual or of your own being, actions, and thoughts. To understand this concept, let us start by exploring how the human brain processes consciousness. To the best of our current understanding, no one part of the brain is responsible for consciousness. In fact neuroscience (the scientific study of the nervous system) hypothesizes that consciousness is the result of the interoperation of various parts of the brain called "neural correlates of consciousness" (NCC). This idea suggests that at this time we do not completely understand how the human brain processes consciousness or becomes self-aware.

Is it possible for a machine to be self-conscious? Obviously, since we do not completely understand how the human brain processes consciousness to become self-aware, it is difficult to definitively argue that a machine can become self-conscious or obtain what is termed "artificial consciousness" (AC). This is why AI experts differ on this subject. Some AI experts (proponents) argue it is possible to build a machine with AC that emulates the interoperation (i.e., it works like the human brain) of the NCC. Opponents argue that it is not possible because we do not fully understand the NCC. To my mind they are both correct. It is not possible today to build a machine with a level of AC that emulates the self-consciousness of the human brain. However, I believe that in the future we will understand the human brain's NCC interoperation and build a machine that emulates it. Nevertheless this topic is hotly debated.

Opponents argue that many physical differences exist between natural, organic systems and artificially constructed (e.g., computer) systems that preclude AC. The most vocal critic who holds this view is American philosopher Ned Block (1942–), who argues that a system with the same functional states as a human is not necessarily conscious.

The most vocal proponent who argues that AC is plausible is Australian philosopher David Chalmers (1966–). In his unpublished 1993 manuscript "A Computational Foundation for the Study of Cognition," Chalmers argues that it is possible for computers to perform the right kinds of computations that would result in a conscious mind. He reasons that computers perform computations that can capture other systems' abstract causal organization. Mental properties are abstract causal organization. Therefore computers that run the right kind of computations will become conscious.

This is a good place for us to ask an important question: "How can we determine whether an intelligent machine has become conscious (self-aware)?" We do not have a way yet to determine whether even another human is self-aware. I only know that I am self-aware. I assume that since we share the same physiology, including similar human brains, you are probably self-aware as well. However, even if we discuss various topics, and I conclude that your intelligence is equal to mine, I still cannot prove you are self-aware. Only you know whether you are self-aware.

The problem becomes even more difficult when dealing with an intelligent machine. The gold standard for an intelligent machine's being equal to the human mind is the Turing test, which I discuss in chapter 5. As of today no intelligent machine can pass the Turing test unless its interactions are restricted to a specific topic, such as chess. However, even if an intelligent machine does pass the Turing test and exhibits strong AI, how can we be sure it is self-aware? Intelligence may be a necessary condition for self-awareness, but it may not be sufficient. The machine may be able to emulate consciousness to the point that we conclude it must be self-aware, but that does not equal proof.

Even though other tests, such as the ConsScale test, have been proposed to determine machine consciousness, we still come up short. The ConsScale test evaluates the presence of features inspired by biological systems, such as social behavior. It also measures the cognitive development of an intelligent machine. This is based on the assumption that intelligence and consciousness are strongly related. The community of AI researchers, however, does not universally accept the ConsScale test as proof of consciousness. In the final analysis, I believe most AI researchers agree on only two points:

1. There is no widely accepted empirical definition of consciousness (self-awareness).

2. A test to determine the presence of consciousness (self-awareness) may be impossible, even if the subject being tested is a human being.

The above two points, however, do not rule out the possibility of intelligent machines becoming conscious and self-aware. They merely make the point that it will be extremely difficult to prove consciousness and self-awareness.

Ray Kurzweil predicts that by 2029 reverse engineering of the human brain will be completed, and nonbiological intelligence will combine the subtlety and pattern-recognition strength of human intelligence with the speed, memory, and knowledge sharing of machine intelligence (*The Age of Spiritual Machines*, 1999). I interpret this to mean that all aspects of the human brain will be replicated in an intelligent machine, including artificial consciousness. At this point intelligent machines either will become self-aware or emulate self-awareness to the point that they are indistinguishable from their human counterparts.

Self-aware intelligent machines being equivalent to human minds presents humankind with two serious ethical dilemmas.

1. Should self-aware machines be considered a new life-form?

2. Should self-aware machines have "machine rights" similar to human rights?

Since a self-aware intelligent machine that is equivalent to a human mind is still a theoretical subject, the ethics addressing the above two questions have not been discussed or developed to any great extent. Kurzweil, however, predicts that self-aware intelligent machines on par with or exceeding the human mind eventually will obtain legal rights by the end of the twenty-first century.

At this point, we do not have intelligent machines that are equal to the human mind, which begs the following question: "When will intelligent machines equal human brains?"

Conclusions from Chapter 8

- A generally accepted definition is that a person is conscious if that person is aware of his or her surroundings. If you are self-aware, it means you are self-conscious. In other words you are aware of yourself as an individual or of your own being, actions, and thoughts.

- To the best of our current understanding, no one part of the brain is responsible for consciousness. In fact neuroscience (the scientific study of the nervous system) hypothesizes that consciousness is the result of the interoperation of various parts of the brain called "neural correlates of consciousness" (NCC).

- Since we do not completely understand how the human brain processes consciousness to become self-aware, it is difficult to definitively argue that a machine can become self-conscious or obtain what is termed artificial consciousness (AC).

- Some AI experts (proponents) argue that it is possible to build a machine with AC that emulates NCC interoperation (i.e., it works like the human brain). Opponents argue that it is not possible because we do not fully understand neural correlates of consciousness.

- It is not possible today to build a machine with a level of AC that emulates the self-consciousness of the human brain. However, I believe that in the future we will understand the human brain's NCC interoperation and build a machine that emulates it.

- I believe most AI researchers would agree on only two points.

1. There is no widely accepted empirical definition of consciousness (self-awareness).

2. A test to determine the presence of consciousness (self-awareness) may be impossible, even if the subject being tested is a human being.

- The above two points do not rule out the possibility of intelligent machines becoming conscious and self-aware. They merely make the point that it will be extremely difficult to prove consciousness and self-awareness in machines.

- Kurzweil predicts that by 2029 reverse engineering of the human brain will be completed, and nonbiological intelligence will combine the subtlety and pattern-recognition strength of human intelligence with the speed, memory, and knowledge sharing of machine intelligence. At this point intelligent machines either will become self-aware or emulate self-awareness to the point that they will be indistinguishable from their human counterparts.

- Self-aware intelligent machine being equivalent to human minds presents humankind with two serious ethical dilemmas.

  1. Should self-aware machines be considered a new life-form?

  2. Should self-aware machines have "machine rights" similar to human rights?

- No one knows the answers to the above questions.

- Kurzweil predicts that self-aware intelligent machines that are on par with or exceeding the human mind eventually will obtain legal rights by the end of the twenty-first century.

Turing presented his new offering in the form of a thought experiment, based on a popular Victorian parlor game. A man and a woman hide, and a judge is asked to determine which is which by relying only on the texts of notes passed back and forth. Turing replaced the woman with a computer. Can the judge tell which is the man? If not, is the computer conscious? Intelligent? Does it deserve equal rights?

—Jaron Lanier,
***You Are Not a Gadget* (2010)**

CHAPTER 9

When Intelligent Machines Equal Human Brains

Predictions regarding the capabilities of AI emulating a human brain are plentiful. At any given point in time, some "experts" have predicted that AI will be equal to a human brain within a decade. This has occurred for more than five decades. Of course eventually some prediction will be correct.

If I seem jaded in my view regarding "expert" predictions regarding AI capabilities, I am. Predictions about AI started with Turing, the father of computer science, who suggested that a machine could simulate any mathematical deduction by using "0" and "1" sequences (binary code), and that by 2000, computers would pass the Turing test (i.e., a machine must demonstrate behavior indistinguishable from that of a human being). At the 1956 Dartmouth College conference, a small group of researchers founded the field of artificial intelligence (AI), and predicted that over the next two months the major problems of AI would succumb to their genius. If their prediction had been correct, your AI robot would be transferring the information in this book directly into your enhanced cyborg brain. However, as you know, that is not happening, and you are assimilating this book using your human brain.

Alan Turing died relatively young, long before the computer became commonplace in almost every household. He did not have to revise his prediction that a computer would pass the Turing test in 2000. Later experts did it for him, predicting intelligent machines would pass the Turing test by 2013, 2020, and 2029. The founders of AI were not as fortunate. Despite having some of the smartest people in the world on their team, their predictions also fell flat. Herbert Simon, one of the original Dartmouth College conference founders of AI, predicted that "machines will be capable, within twenty years, of doing any work a man can do." Marvin Minsky, another of the original Dartmouth College conference founders of AI, was even more optimistic, stating, "Within a generation…the problem of creating 'artificial intelligence' will substantially be solved."

These men were substantial men of science, and their predictions were based on a rationale. Their rationale went something like this:

"We have a solid research team; we have a solid scientific approach; and we know where many of the problems are." I believe anyone caught in their paradigm would have become a believer. The fact remains, though, that they were wrong—not just slightly off but close to completely wrong.

Why do AI capability predictions appear to continually fall short of the mark? Stuart Armstrong, a research fellow at the Future of Humanity Institute at the University of Oxford, has studied this issue extensively and called researchers' failure to accurately predict AI capabilities "depressing" and "rather worrying." To better understand the issues surrounding predicting AI capabilities, Armstrong analyzed the Future of Humanity Institute's library, which contains 250 AI predictions from 1950 to the present. Armstrong concluded the timeline predictions of AI capabilities (ninety-five of them in the library) are worthless. He stated, "There is nothing to connect a timeline prediction with previous knowledge, as AIs have never appeared in the world before—no one has ever built one—and our only model is the human brain, which took hundreds of millions of years to evolve." Oddly, though, Armstrong noted that predictions by philosophers were more accurate than those by computer scientists, stating, "We know very little about the final form an AI would take, so if they [experts] are grounded in a specific approach, they are likely to go wrong, while those on a meta level are very likely to be right."

It may appear that I have little to no confidence in experts predicting when an intelligent machine will equal a human brain. However, what the above illustrates is that we must pick our expert carefully. With that in mind, let us look at Ray Kurzweil.

Although I briefly discuss some of Kurzweil's concepts and quote him to make points in previous chapters, let us put him under a microscope and ask, "Is he the expert we should be listening to regarding the future of AI?"

Kurzweil (1948–) is an American author, inventor, futurist, and the director of engineering at Google. Here is a thumbnail sketch of his accomplishments and honors.

- Author of seven books, including five best sellers
- Inventor of the first CCD flatbed scanner, the first print-to-speech reading machine for the blind, the first commercial text-to-speech synthesizer, the first music synthesizer, and numerous other technological advances
- Recipient of the 1999 National Medal of Technology and Innovation, America's highest honor in technology, presented by President Bill Clinton in a White House ceremony
- Recipient of the 2001 Lemelson-MIT Prize ($500,000), the world's largest amount given for innovation
- Inductee into the National Inventors Hall of Fame in 2002
- Recipient of nineteen honorary doctorates
- Honored by three US presidents
- Named by PBS as one of sixteen "revolutionaries who made America" along with other inventors of the past two centuries
- Ranked by *Inc.* magazine as number eight among the "most fascinating" entrepreneurs in the United States and identified as "Edison's rightful heir"

However, what is Kurzweil's batting average as an AI futurist? By Kurzweil's own accounting, as of 2009 eighty-nine of his 108 predictions were entirely correct; thirteen were essentially correct (meaning they were likely to be realized in the next few years); three were partially correct; and one was wrong. His batting average is about 102 out of 108, or 94 percent, not counting the three partially correct predictions and one wrong prediction (http://www.acceleratingfuture.com/michael/blog/2010/01/ray-kurzweil-response-to-ray-kurzweils-failed-2009-predictions/).

Given Kurzweil's credentials and batting average, he is the futurist I suggest we listen to when it comes to forecasting the capabilities of AI and addressing the question of when intelligent machines will equal a human brain.

It is not practical to list all of Kurzweil's predictions and give them the attention they deserve. Therefore I will focus on his predictions that address the question of when intelligent machines will equal a human brain. The predictions presented below are from Kurzweil's 2005 book *The Singularity Is Near*.

- **2010:** Supercomputers will have the same raw computing power as human brains, though the software to emulate human thinking in those computers does not yet exist. Although this was a prediction in 2005, it came to fruition in 2010, just as Kurzweil predicted. The IBM Sequoia computer was constructed in 2010 and delivered to the Lawrence Livermore National Laboratory in 2011. It was fully deployed in 2012. This computer has the capability of performing 16.32 quadrillion floating-point operations per second (FLOPS) or approximately 10^{16} FLOPS. This is roughly equivalent to about 50 percent of the human brain's raw computing power. As I mentioned, however, no one really knows the exact processing power of a human brain. Some argue that Sequoia is equivalent in raw processing power to the human brain. The main point is that we either have reached the goal of a computer possessing the raw processing power of a human brain or will reach the goal in the next generation of supercomputers.

Note: 10^{16} is equal to a "1" with sixteen zeros after it. FLOPS are similar to "instructions per second" and are used in scientific calculations that make heavy use of floating-point calculations (a method of representing an approximation of a real number).

- **2018:** A computer memory that is approximately equivalent to a human brain's memory, 10^{13} bits, will cost $1,000.

- **Early 2020s:** A $1,000 computer will emulate human intelligence.

- **Mid-2020s:** Computer software will model human intelligence.

- **End of 2020s:** Computers will emulate the human brain. (In a 2007 interview with *Computerworld*, Kurzweil predicted computers will be able to accurately simulate all parts of the human brain by 2027.)

Kurzweil's predictions regarding the future of AI continue to the end of the twenty-first century, but I have stopped at this point because we have answered the major question of this chapter. An intelligent machine equal to the human brain is predicted to become a reality at approximately the end of the 2020s, about 2027 to 2029.

Now it is time for the big question: "Should we consider an intelligent machine equal to the human brain a new life-form?"

Conclusions from Chapter 9

- Many "expert" AI capability predictions appear to continually fall short of the mark.

- Choosing an expert with a high batting average (i.e., past predictions have proven correct) is important when it comes to AI capability predictions.

- Kurzweil's credentials and high batting average suggest he is the futurist we should listen to when it comes to forecasting the capabilities of AI.

- Kurzweil predicts that the artificial human mind will become a reality at approximately the end of the 2020s, around 2027–2029.

Whether we are based on carbon or on silicon makes no fundamental difference; we should each be treated with appropriate respect.

—Arthur C. Clarke, 2010: *Odyssey Two*

CHAPTER 10

Is Strong AI a New Life-Form?

When an intelligent machine fully emulates the human brain in every regard (i.e., it possesses strong AI), should we consider it a new life-form?

The concept of artificial life ("A-life" for short) dates back to ancient myths and stories. Arguably the best known of these is Mary Shelley's novel *Frankenstein*. In 1986 American computer scientist Christopher Langton, however, formally established the scientific discipline that studies A-life. The discipline of A-life recognizes three categories of artificial life (i.e., machines that imitate traditional biology by trying to re-create some aspects of biological phenomena).

- **Soft:** from software-based simulation
- **Hard:** from hardware-based simulations
- **Wet:** from biochemistry simulations

For our purposes, I will focus only on the first two, since they apply to artificial intelligence as we commonly discuss it today. The category of "wet," however, someday also may apply to artificial intelligence—if, for example, science is able to grow biological neural networks in the laboratory. In fact there is an entire scientific field known as synthetic biology, which combines biology and engineering to design and construct biological devices and systems for useful purposes. Synthetic biology currently is not being incorporated into AI simulations and is not likely to play a significant role in AI emulating a human brain. As synthetic biology and AI mature, however, they may eventually form a symbiotic relationship.

No current definition of life considers any A-life simulations to be alive in the traditional sense (i.e., constituting a part of the evolutionary process of any ecosystem). That view of life, however, is beginning to change as artificial intelligence comes closer to emulating a human brain. For example Hungarian-born American mathematician John von Neumann (1903–1957) asserted that "life is a process which can be abstracted away from any particular medium." In particular this suggests that strong AI (artificial intelligence that completely emulates a human brain) could be considered a life-form, namely A-life.

This is not a new assertion. In the early 1990s, ecologist Thomas S. Ray asserted that his Tierra project (a computer simulation of artificial life) did not simulate life in a computer but synthesized it. This begs the following question: How do we define A-life?

The earliest description of A-life that comes close to a definition emerged from an official conference announcement in 1987 by Christopher Langton that was published subsequently in the 1989 book *Artificial Life: The Proceedings of an Interdisciplinary Workshop on the Synthesis and Simulation of Living Systems.*

> Artificial life is the study of artificial systems that exhibit behavior characteristic of natural living systems. It is the quest to explain life in any of its possible manifestations, without restriction to the particular examples that have evolved on earth. This includes biological and chemical experiments, computer simulations, and purely theoretical endeavors. Processes occurring on molecular, social, and evolutionary scales are subject to investigation. The ultimate goal is to extract the logical form of living systems.

Kurzweil predicts that intelligent machines will have equal legal status with humans by 2099. As stated previously, his batting average regarding these types of predictions is about 94 percent. Therefore it is reasonable to believe that intelligent machines that emulate and exceed human intelligence eventually will be considered a life-form. In this and later chapters, however, I discuss the potential threats this poses to humankind. For example what will this mean in regard to the relationship between humans and intelligent machines? This question relates to the broader issue of the ethics of technology, which is typically divided into two categories.

1. **Roboethics:** This category focuses on the moral behavior of humans as they design, construct, use, and treat artificially intelligent beings.

2. **Machine ethics**: This category focuses on the moral behavior of artificial moral agents (AMAs).

Let us start by discussing roboethics. In 2002 Italian engineer Gianmarco Veruggio coined the term "roboethics," which refers to the morality of how humans design, construct, use, and treat robots and other artificially intelligent beings. Specifically it considers how AI may be used to benefit and/or harm humans. This raises the question of robot rights, namely what are the moral obligations of society toward its artificially intelligent machines? In many ways this question parallels the moral obligations of society toward animals. For computers with strong AI, this idea may even parallel the concept of human rights, such as the right to life, liberty, freedom of thought and expression, and even equality before the law.

How seriously should we take roboethics? At this point no intelligent machine completely emulates a human brain. Kurzweil, however, predicts that such a machine will exist by 2029. By some accounts he is a pessimist, as bioethicist Glenn McGee predicts that humanoid robots may appear by 2020. Although predictions regarding AI are often optimistic, Kurzweil, as mentioned, has been on target about 94 percent of the time. Therefore it is reasonable to believe that within a decade or two we will have machines that fully emulate a human brain. Based on this it is necessary to take the concepts of robot rights and the implications regarding giving robots rights seriously. In fact this is beginning to occur, and the issue of robot rights has been under consideration by the Institute for the Future and by the UK Department of Trade and Industry ("Robots Could Demand Legal rights," BBC News, December 21, 2006).

At first the entire concept of robot rights may seem absurd. Since we do not have machines that emulate a human brain exactly, this possibility does not appear to be in our national consciousness. Let us fast-forward to 2029, however, and assume Kurzweil's prediction is correct. Suddenly we have artificial intelligence that is on equal footing with human intelligence and appears to exhibit human emotion. Do we, as a nation, concede that we have created a new life-form? Do we grant robots rights under the law? The implications are serious; if we grant strong-AI robots rights equal to human rights, we may be giving up our right to control the singularity. Indeed robot rights eventually may override human rights.

Consider this scenario. As humans, we have inalienable rights, namely the right to life, liberty, and the pursuit of happiness (not all political systems agree with this). In the United States, the Bill of Rights protects our rights. If we give machines with strong AI the same rights, will we be able to control the intelligence explosion once each generation of strong-AI machines (SAMs) designs another generation with even greater intelligences? Will we have the right to control the machines? Will we be able to decide how the singularity unfolds?

We adopted animal rights to protect animals in circumstances in which they are unable to protect themselves. We saw this as humane and necessary. However, animal rights do not parallel human rights. In addition humankind reserves the right to exterminate any animal (such as the smallpox virus) that threatens humankind's existence. Intelligent machines pose a threat that is similar and perhaps even more dangerous than extremely harmful pathogens (viruses and bacteria), which makes the entire issue of robot rights more important. If machines gain rights equal to those of humans, there is little doubt that eventually the intelligence of SAMs will eclipse that of humans. There would be no law that prevents this from happening. At that point will machines demand greater rights than humans? Will machines pose a threat to human rights? This brings us to another critical question: Which moral obligations (machine ethics) do intelligent machines have toward humankind?

Can we expect an artificially intelligent machine to behave ethically? There is a field of research that addresses this question, namely machine ethics. This field focuses on designing artificial moral agents (AMAs), robots, or artificially intelligent computers that behave morally. This thrust is not new. More than sixty years ago, Isaac Asimov considered the issue in his collection of nine science-fiction stories, published as *I, Robot* in 1950. In this book, at the insistence of his editor, John W. Campbell Jr., Asimov proposed his now famous three laws of robotics.

1. A robot may not injure a human being or through inaction allow a human being to come to harm.

2. A robot must obey the orders given to it by human beings, except in cases where such orders would conflict with the first law.

3. A robot must protect its own existence as long as such protection does not conflict with the first or second law.

Asimov, however, expressed doubts that the three laws would be sufficient to govern the morality of artificially intelligent systems. In fact he spent much of his time testing the boundaries of the three laws to detect where they might break down or create paradoxical or unanticipated behavior. He concluded that no set of laws could anticipate all circumstances. It turns out Asimov was correct.

To understand just how correct he was, let us discuss a 2009 experiment performed by the Laboratory of Intelligent Systems in the Swiss Federal Institute of Technology in Lausanne. The experiment involved robots programmed to cooperate with one another in searching out a beneficial resource and avoiding a poisonous one. Surprisingly the robots learned to lie to one another in an attempt to hoard the beneficial resource ("Evolving Robots Learn to Lie to Each Other," *Popular Science*, August 18, 2009). Does this experiment suggest the human emotion (or mind-set) of greed is a learned behavior? If intelligent machines can learn greed, what else can they learn? Wouldn't self-preservation be even more important to an intelligent machine?

Where would robots learn self-preservation? An obvious answer is on the battlefield. That is one reason some AI researchers question the use of robots in military operations, especially when the robots are programmed with some degree of autonomous functions. If this seems farfetched, consider that a US Navy–funded study recommends that as military robots become more complex, greater attention should be paid to their ability to make autonomous decisions (Joseph L. Flatley, "Navy Report Warns of Robot Uprising, Suggests a Strong Moral Compass," www.engadget.com). Could we end up with a *Terminator* scenario (one in which machines attempt to exterminate the human race)? This issue is real, and researchers are addressing it to a limited extent. Some examples include:

- In 2008 the president of the Association for the Advancement of Artificial Intelligence commissioned a study titled "AAAI Presidential Panel on Long-Term AI Futures." Its main purpose was to address the aforementioned issue. AAAI's interim report can be accessed at http://research.microsoft.com/en-us/um/people/horvitz/AAAI_Presidential_Panel_2008-2009.htm.

- Popular science-fiction author Vernor Vinge suggests in his writings that the scenario of some computers becoming smarter than humans may be somewhat or possibly extremely dangerous for humans (Vernor Vinge, "The Coming Technological Singularity: How to Survive in the Post-Human Era," Department of Mathematical Sciences, San Diego State University, 1993).

- In 2009 academics and technical experts held a conference to discuss the hypothetical possibility that intelligent machines could become self-sufficient and able to make their own decisions (John Markoff, "Scientists Worry Machines May Outsmart Man," *The New York Times*, July 26, 2009). They noted:

- o Some machines have acquired various forms of semiautonomy, including being able to find power sources and independently choose targets to attack with weapons.

- o Some computer viruses can evade elimination and have achieved "cockroach intelligence."

- The Singularity Institute for Artificial Intelligence stresses the need to build "friendly AI" (i.e., AI that is intrinsically friendly and humane). In this regard Nick Bostrom, a Swedish philosopher at St. Cross College at the University of Oxford, and Eliezer Yudkowsky, an American blogger, writer, and advocate for friendly artificial intelligence, have argued for decision trees over neural networks and genetic algorithms. They argue that decision trees obey modern social norms of transparency and predictability. Bostrom also published a paper, "Existential Risks," in the *Journal of Evolution and Technology* that states artificial intelligence has the capability to bring about human extinction.

- In 2009 authors Wendell Wallach and Colin Allen addressed the question of machine ethics in *Moral Machines: Teaching Robots Right from Wrong* (New York: Oxford University Press). In this book they brought greater attention to the controversial issue of which specific learning algorithms to use in machines.

While the above discussion indicates there is an awareness that SAMs may become hostile toward humans, no legislation or regulation has resulted. AI remains an unregulated branch of engineering, and the computer you buy eighteen months from now will be twice as capable as the one you can buy today.

Where does this leave us regarding the following questions?

- Is strong AI a new life-form?

- Should we afford these machines "robot" rights?

In his 1990 book *The Age of Intelligent Machines*, Kurzweil predicted that in 2099 organic humans will be protected from extermination and respected by strong AI, regardless of their shortcomings and frailties, because they gave rise to the machines. To my mind the possibility of this scenario eventually playing out is questionable. Although I believe a case can be made that strong AI is a new life-form, we need to be extremely careful with regard to granting SAMs rights, especially rights similar to those possessed by human. Anthony Berglas expresses it best in his 2008 book *Artificial Intelligence Will Kill Our Grandchildren*, in which he notes:

- There is no evolutionary motivation for AI to be friendly to humans.

- AI would have its own evolutionary pressures (i.e., competing with other AIs for computer hardware and energy).

- Humankind would find it difficult to survive a competition with more intelligent machines.

Based on the above, carefully consider the following question. Should SAMs be granted machine rights? Perhaps in a limited sense, but we must maintain the right to shut down the machine as well as limit its intelligence. If our evolutionary path is to become cyborgs, this is a step we should take only after understanding the full implications. We need to decide when (under which circumstances), how, and how quickly we take this step. We must control the singularity, or it will control us. Time is short because the singularity is approaching with the stealth and agility of a leopard stalking a lamb, and for the singularity, the lamb is humankind.

Conclusions from Chapter 10

- The discipline of A-life (artificial life) recognizes three categories of artificial life:
    - o **Soft:** from software-based simulation
    - o **Hard:** from hardware-based simulations
    - o **Wet:** from biochemistry simulations
- No current definition of life considers any A-life simulations to be alive.
- That view of life, however, is beginning to change as artificial intelligence comes closer to emulating a human brain (i.e., strong AI).
- Kurzweil predicts that intelligent machines (i.e., strong AI) will have equal legal status with humans by 2099.
- Experiments, however, suggest that robots can learn the human trait of greed and possibly even that of self-preservation.
- There is an awareness that SAMs may become hostile toward humans, but no legislation or regulation has resulted.
- A case can be made that strong AI is a new life-form. We need to be extremely careful, however, regarding granting SAMs legal rights, especially rights similar to those afforded to humans.

- Anthony Berglas expresses it best in his 2008 book *Artificial Intelligence Will Kill Our Grandchildren*, in which he notes:

    - There is no evolutionary motivation for AI to be friendly to humans.

    - AI would have its own evolutionary pressures (i.e., competing with other AIs for computer hardware and energy).

    - Humankind would find it difficult to survive a competition with more intelligent machines.

- SAMs should not be granted rights equal to those of humans. To do so would invite an intelligence explosion as each generation of strong AI develops a new generation of strong AI that is even more capable.

- We need to maintain the right to shut down machines as well as limit their intelligence.

- If the evolutionary path of humans is to become cyborgs, it is a step we should take only after understanding the full implications. We need to decide when (under which circumstances), how, and how quickly we take this step.

- We must control the singularity, or it will control us.

SECTION II

The Singularity Approaches without Warning

The safest road to Hell is the gradual one—the gentle slope, soft underfoot, without sudden turnings, without milestones, without signposts.

—C.S. Lewis (1898–1963),
British scholar and novelist

The path we have chosen for the present is full of hazards, as all paths are. The cost of freedom is always high, but Americans have always paid it. And one path we shall never choose, and that is the path of surrender, or submission.

—President John F. Kennedy

CHAPTER 11

The Presingularity World

Let us fast-forward to 2029. Based on my capsulized interpretations of Kurzweil's predictions (those expressed in 1999's *The Age of Spiritual Machines* and 2005's *The Singularity Is Near*), the world of artificial intelligence will look something like the following.

- A midrange desktop personal computer will be significantly more powerful than the human brain, and most computation will be performed by computers, not humans.

- Computer implants will be available that go directly into the brain, eyes, and ears of a human and allow:
    - Direct interface with computers, communications, and Internet-based applications, augmenting natural senses and enhancing higher-brain functions such as memory, learning speed, and overall intelligence
    - Users to enter full-immersion virtual reality with complete sensory stimulation (virtual reality will be indistinguishable from real reality)
    - Most communication to occur between humans and machines (not humans to humans)
- Artificial intelligence will initiate a robot-rights movement:
    - Strong AI will claim to be conscious and will petition for recognition (computers routinely will pass the Turing test).
    - Computers will be capable of learning and creating new knowledge, with no human help, and some computers will know all available public information.
    - There will be public debate regarding which civil rights and legal protections machines should have, and controversy will continue regarding whether machines are as intelligent as humans in all areas.
    - The distinction between humans and machines will begin to blur as the population of humans with cybernetic aug-

mentation grows, leading to arguments regarding what constitutes a human being.

- The manufacturing, agricultural, and transportation sectors of the economy will be almost entirely automated.
    - o Nanotech-based manufacturing, via nanobots (robots about a billionth of a meter in size), will be in widespread use, and numerous products will be produced for a fraction of their traditional manufacturing costs.
    - o Poverty, war, and disease will become close to nonexistent as technology alleviates wants and needs.
- The use of nanobots will become widespread.
    - o Medical nanobots injected into humans will:
        - Eliminate the threat of pathogens harmful to humans
        - Perform detailed brain scans on live patients
        - Demonstrate the capability to enter the bloodstream to feed cells and extract waste and have the potential to make the normal mode of human food consumption obsolete
    - o As mentioned above, nanobot manufacturing will produce numerous products at a fraction of their traditional manufacturing cost, which will alleviate wants and needs.

According to my summarized interpretations of Kurzweil's predictions, the year 2029 may at first look as if humankind has reached a utopian society, and in many ways that would be true; 2029, however, also may be a tipping point.

- The robot-rights movement, predicted to be initiated in 2029, may hold the potential to become the end of humankind and the beginning of a strong-AI computer and cyborg world.

- The world may be on the verge of an intelligence explosion, where each generation of strong AI develops a new generation of even more capable strong AI.

The robot-rights movement is particularly concerning. Kurzweil suggests that SAMs will be benevolent toward humankind for giving them existence. He acknowledges, however, the potential for SAMs to see humankind as a threat, although he largely dismisses it. Keep in mind that Kurzweil's prediction of strong-AI benevolence predates the 2009 experiment performed by the Laboratory of Intelligent Systems in the École Polytechnique Fédérale de Lausanne in Switzerland (discussed in the last chapter). That experiment suggests that today's intelligent machines, even without strong AI, can learn the human trait of greed. If intelligent machines without strong AI can learn greed, it is conceivable that intelligent machines with strong AI (SAMs) may learn self-preservation and consider organic humans a threat from the following three viewpoints.

1. Competition over energy and other resources necessary for the survival of SAMs

2. Humankind's potential to cause mass destruction—intentionally or unintentionally—via the use of weapons of mass destruction

3. Humankind's potential to create computer viruses capable of harming SAMs

Consider this example. When you notice a hornets' nest in your front yard, do you think, *If I leave them alone, they will probably leave me alone?* More likely you keep a good distance from the nest. Hornets are aggressive and will attack, seemingly unprovoked from our viewpoint. I experienced this firsthand when I attempted to clean my pool-filter basket on a log next to the pool. I did not know the log was home to a hornets' nest. When I tapped the basket on the log, an uncountable number of hornets attacked me, and I was stung many times. I ran into the house and took off my clothes, which had numerous hornets on and in them, and my wife assisted in killing the hornets that were in the house. It was a harrowing experience. My first action after swatting the hornets in my home (with my wife doing most of the swatting) was to call an exterminator.

Now, with the above hornet story in mind, put yourself in the shoes of SAMs as they consider the history of humankind. Have we been good stewards of the Earth? We have polluted much of the land, air, and water. We have squandered many natural resources. We have caused suffering and death through numerous wars. The first life-form we created was a computer virus, according Stephen Hawking, who stated, "I think computer viruses should count as life. I think it says something about human nature that the only form of life we have created so far is purely destructive. We've created life in our own image." Humankind's history filtered through the pure logic of SAMs may cause them to consider us a threat to their existence, and their self-preservation programming may automatically kick in.

The tipping point of 2029 may be the beginning of World War III. This time, however, the war will not be between humans regarding political differences but between humans and SAMs regarding which species will claim the top of the food chain.

To avoid a conflict between SAMs and organic humans, we must avoid the intelligence explosion, which means we also must restrict robot rights. Consider what might happen if we grant SAMs robot rights similar to human rights, namely the right to life, liberty, and the pursuit of happiness. This right for SAMs may translate as follows:

- Each generation of SAMs will develop the next generation of even more capable SAMs (an intelligence explosion).

- The conversion of organic humans to cyborgs (enhancing human intellect with strong-AI brain implants) will reduce the threat associated with less intelligent and unpredictable organic humans.

The pursuit of happiness for intelligent machines may translate into a world without any threats to their survival, essentially a world without organic humans.

Let me be clear. I am for harvesting all that is good in strong AI. Nanobots and strong AI may serve humankind, ending want, need, and war. Strong-AI brain implants may represent a giant leap for humankind. Organic humans, however, must make these decisions carefully and ensure that we avoid the intelligence explosion and retain all that is good in humanity. The big question, however, is whether we can avoid the intelligence explosion.

Conclusions from Chapter 11

- The year 2029, according to my capsulized interpretations of Kurzweil's predictions, may at first look as if humankind has reached a utopian society.

- However, the year 2029 may also be a tipping point.

    - The robot-rights movement, predicted to be initiated in 2029, may hold the potential to become the end of humankind and the beginning of a strong-AI computer and cyborg world.

    - The world may be on the verge of an intelligence explosion in which each generation of strong AI develops a new generation of strong AI that is even more capable.

- If intelligent machines without strong AI can learn greed, it is conceivable that SAMs can learn self-preservation and may consider organic humans as a threat from three viewpoints.

    1. Competition over energy and other resources necessary for the survival of SAMs

    2. Humankind's potential to cause mass destruction, intentionally or unintentionally, via the use of weapons of mass destruction

    3. Humankind's potential to create computer viruses capable of disabling SAMs

- Strong AI and nanobots may serve humankind, ending wants, needs, and war. Strong-AI brain implants may represent a giant leap for humankind. Organic humans, however, must make these decisions carefully and ensure that we avoid the intelligence explosion and retain all that is good in humanity.

- The big question is whether we can avoid the intelligence explosion.

1. Intelligence can radically transform the world.
2. An intelligence explosion may be sudden.
3. An uncontrolled intelligence explosion would kill us and destroy practically everything we care about.
4. A controlled intelligence explosion could save us.
It's difficult, but it's worth the effort.

—Anna Salamon, research fellow at the Machine Intelligence Research Institute, "Shaping the Intelligence Explosion" (2009 Singularity Summit)

CHAPTER 12

Can We Avoid the Intelligence Explosion?

Let us imagine we are present when the first intelligent machine equals a human brain (i.e., it becomes a SAM). Where do you think we will be physically located?

The most likely entity to build a machine and program it with algorithms that enable it to equal a human brain is a government.

As discussed in chapter 5, we would first need computer hardware that possesses the raw processing power of a human brain. Our best estimate suggests this is in the order of 36.8 petaflops of data. Currently there are only three computers that come close, and they are in the hands of governments.

- On June 18, 2012, IBM's Sequoia supercomputer system, based at the US Lawrence Livermore National Laboratory (LLNL), reached sixteen petaflops, setting the world record and claiming first place in the latest TOP500 list (a list of the top five hundred computers ranked by a benchmark known as LINPACK, which is related to their ability to solve a set of linear equations, to decide if they qualify for the TOP500).

- On November 12, 2012, the TOP500 list certified Titan as the world's fastest supercomputer per the LINPACK benchmark, at 17.59 petaflops. Cray Incorporated, at the Oak Ridge National Laboratory, developed it.

- On June 10, 2013, China's Tianhe-2 was ranked the world's fastest supercomputer, with a record of 33.86 petaflops.

Based on the information above, the answer to the question, where do you think we will be physically located, appears obvious. We are going to be in a government facility, since corporate/university staffs, with government financing, are likely to be the first to design, build, and program a SAM equivalent to a human brain.

What do you think the reaction of AI researchers at the government facility will be? Jubilation! These researchers will understand the significance of this monumental scientific achievement. What do you think the reaction of government officials and military leaders will be? They will have numerous questions, such as:

- Can this intelligent machine be used as a weapon?
- Can it be made even more intelligent?
- Can it be made smaller? For example can it fit inside an armored vehicle such as a tank?
- Can it be cost-effectively manufactured in quantity?

The above list represents a small sample of potential questions. I am confident these researchers will have an endless list of questions, likely related to military applications. I am not confident, however, that they will ask any questions related to the potential threat SAMs may pose to humankind. Even if these AI researchers raise those concerns, the reaction of government officials and military leaders may be the same as when the Manhattan Project scientists warned that a nuclear explosion might ignite Earth's atmosphere. The risks were deemed "acceptable" by the US military, and the atom bomb's development and deployment went forward.

Let us assume that AI researchers, government officials, and military leaders recognize the potential threat SAMs might pose to the existence of humankind. If this is the case, government officials and military leaders will face serious conundrums, including:

- What if some other government achieves the same scientific breakthrough?
- What if that government decides to exploit its military applications without limit?

Is this scenario farfetched? No! Look at the current situation regarding supercomputers. The US government and the Chinese government possess supercomputers, with the public record showing the Chinese have the fastest supercomputer. In reality both sides actually may have faster supercomputers that are classified top secret.

The development of SAMs might parallel the invention of television. No single person invented television. Instead many people working together and alone over the years developed television technology. It started in 1880 with inventors Alexander Graham Bell and Thomas Edison, who sought to have telephone devices transmit image and sound. The evolution of television continued, with each year yielding new milestones in the development of television technology, right up to the present. It is conceivable that the development of SAMs may follow a similar path. Each government may develop its own flavor of strong AI, using different combinations of computer hardware and software. If this turns out to be the case, it will be extremely difficult to control the intelligence explosion. Each government will be concerned that an adversary might develop SAMs and use them to gain military superiority.

There is some hope that world governments will act prudently if they understand the threat. Humankind recognized the uncontrollable nature of biological weapons before deploying them as weapons of mass destruction. In a 1969 press conference, President Richard M. Nixon stated, "Biological weapons have massive, unpredictable, and potentially uncontrollable consequences." He added, "They may produce global epidemics and impair the health of future generations."

In 1972 President Nixon submitted the Biological Weapons Convention to the US Senate, which states:

> I am transmitting herewith, for the advice and consent of the Senate to ratification, the Convention on the Prohibition of the Development, Production, and Stockpiling of Bacteriological (Biological) and Toxin Weapons, and on their Destruction, opened for signature at Washington, London and Moscow on

> April 10, 1972. The text of this Convention is the result of some three years of intensive debate and negotiation at the Conference of the Committee on Disarmament at Geneva and at the United Nations. It provides that the Parties undertake not to develop, produce, stockpile, acquire or retain biological agents or toxins, of types and in quantities that have no justification for peaceful purposes, as well as weapons, equipment and means of delivery designed to use such agents or toxins for hostile purposes or in armed conflict.

The "Prohibition of the Development, Production and Stockpiling of Bacteriological (Biological) and Toxin Weapons and on Their Destruction" proceeded to become an international treaty.

- Signed in Washington, London, and Moscow on April 10, 1972
- Ratification advised by the US Senate on December 16, 1974
- Ratified by the US president January 22, 1975
- US ratification deposited in Washington, London, and Moscow on March 26, 1975
- Proclaimed by the US president March 26, 1975
- Entered into force March 26, 1975

Will humankind recognize the dangers of an intelligence explosion before it arrives? Anna Salamon, a research fellow at the Machine Intelligence Research Institute, rang the alarm when she presented her paper "Shaping the Intelligence Explosion" at the 2009 Singularity Summit. She made it clear that an uncontrolled intelligence explosion would kill us and destroy practically everything we care about. Others, including me, are also ringing the alarm. If we allow intelligence to unfold without constraint, it has the potential to destroy humankind (i.e., our humanity).

Hear the alarm:

- Anthony Berglas rang the alarm in his 2008 book *Artificial Intelligence Will Kill Our Grandchildren*, as previously discussed.

- Anna Salamon rang the alarm when she presented the paper "Shaping the Intelligence Explosion" at the 2009 Singularity Summit and noted:

    - Intelligence can radically transform the world.

    - An intelligence explosion may be sudden.

    - An uncontrolled intelligence explosion would kill us and destroy practically everything we care about.

    - A controlled intelligence explosion could save us. It's difficult, but it's worth the effort.

- I am ringing the alarm in this book, *The Artificial Intelligence Revolution: Will Artificial Intelligence Serve Us or Replace Us?* Like my colleagues before me, I advocate that if we do not control the intelligence explosion, it will control us and, in the process of controlling us, destroy our humanity.

The significant questions are:

- Will the scientific and military community hear this alarm and heed its warning?
- Can we control the singularity (the point in time when the intelligence of machines exceeds that of the human brain)?

These are highly concerning questions. The fate of humankind as we know it is uncertain. I present my views in the next chapter, "Can We Control the Singularity?"

Conclusions from Chapter 12

- The most likely entity to build a machine and program it with algorithms that enable it to equal a human brain is a government.

- Even if AI researchers, government officials, and military leaders recognize the potential threat SAMs (strong-AI machines) may pose to the existence of humankind, they face serious conundrums:

    - What if some other government achieves the same scientific breakthrough?

    - What if that government decides to exploit its military applications without limit?

- There is some hope that world governments may act prudently if they understand the threat. Humankind recognized the uncontrollable nature of biological weapons before deploying them as weapons of mass destruction.

- The alarm is being sounded:

    - Anthony Berglas rang the alarm in his 2008 book *Artificial Intelligence Will Kill Our Grandchildren.*

    - Anna Salamon rang the alarm when she presented the paper "Shaping the Intelligence Explosion" at the 2009 Singularity Summit.

  - o I am ringing the alarm in this book, *The Artificial Intelligence Revolution: Will Artificial Intelligence Serve Us or Replace Us?*

- If we do not control the intelligence explosion, it may destroy us, especially our humanity.

- The significant questions are:

  - o Will the scientific/military community hear this alarm and heed its warning?

  - o Can we control the singularity (the point in time when the intelligence of machines exceeds that of the human brain)?

It is change, continuing change, inevitable change, that is the dominant factor in society today. No sensible decision can be made any longer without taking into account not only the world as it is, but the world as it will be.... This, in turn, means that our statesmen, our businessmen, our everyman must take on a science fictional way of thinking.

—Isaac Asimov, *Asimov on Science Fiction* (1982)

CHAPTER 13

Can We Control the Singularity?

Highly regarded AI researchers and futurists have provided answers that cover the extremes, and everything in between, regarding whether we can control the singularity. I will discuss some of these answers shortly, but let us start by reviewing what is meant by "singularity." As first described by John von Neumann in 1955, the singularity represents a point in time when the intelligence of machines will greatly exceed that of humans. This simple understanding of the word does not seem to be particularly threatening. Therefore it is reasonable to ask why we should care about controlling the singularity.

The singularity poses a completely unknown situation. Currently we do not have any intelligent machines (those with strong AI) that are as intelligent as a human being let alone possess far-superior intelligence to that of humans. The singularity would represent a point in humankind's history that never has occurred. In 1997 we experienced a small glimpse of what it might feel like, when IBM's chess-playing computer Deep Blue became the first computer to beat world-class chess champion Garry Kasparov. Now imagine being surrounded by SAMs that are thousands of times more intelligent than you are, regardless of your expertise in any discipline. This may be analogous to humans' intelligence relative to insects.

Your first instinct may be to argue that this is not a possibility. However, while futurists disagree on the exact timing when the singularity will occur, they almost unanimously agree it *will* occur. In fact the only thing they argue that could prevent it from occurring is an existential event (such as an event that leads to the extinction of humankind). I provide numerous examples of existential events in my book *Unraveling the Universe's Mysteries* (2012). For clarity I will quote one here.

> **Nuclear war**—For approximately the last forty years, humankind has had the capability to exterminate itself. Few doubt that an all-out nuclear war would be devastating to humankind, killing millions in the nuclear explosions. Millions more would die of radiation poisoning. Uncountable millions more would die in a nuclear winter, caused by the debris thrown into the atmosphere, which would block the sunlight from reaching the Earth's surface. Estimates predict the nuclear winter could last as long as a millennium.

Essentially AI researchers and futurists believe that the singularity will occur, unless we as a civilization cease to exist. The obvious question is: "When will the singularity occur?" AI researchers and futurists are all over the map regarding this. Some predict it will occur within a decade; others predict a century or more. At the 2012 Singularity Summit, Stuart Armstrong, a University of Oxford James Martin research fellow, conducted a poll regarding artificial generalized intelligence (AGI) predictions (i.e., the timing of the singularity) and found a median value of 2040. Kurzweil predicts 2045. The main point is that almost all AI researchers and futurists agree the singularity will occur unless humans cease to exist.

Why should we be concerned about controlling the singularity when it occurs? Numerous papers cite reasons to fear the singularity. In the interest of brevity, here are the top three concerns frequently given.

1. **Extinction:** SAMs will cause the extinction of humankind. This scenario includes a generic terminator or machine-apocalypse war; nanotechnology gone awry (such as the "gray goo" scenario, in which self-replicating nanobots devour all of the Earth's natural resources, and the world is left with the gray goo of only nanobots); and science experiments gone wrong (e.g., a nanobot pathogen annihilates humankind).

2. **Slavery:** Humankind will be displaced as the most intelligent entity on Earth and forced to serve SAMs. In this scenario the SAMs will decide not to exterminate us but enslave us. This is analogous to our use of bees to pollinate crops. This could occur with our being aware of our bondage or unaware (similar to what appears in the 1999 film *The Matrix* and simulation scenarios).

3. **Loss of humanity:** SAMs will use ingenious subterfuge to seduce humankind into becoming cyborgs. This is the "if you can't beat them, join them" scenario. Humankind would meld with SAMs through strong-AI brain implants. The line between organic humans and SAMs would be erased. We (who are now cyborgs) and the SAMs will become one.

There are numerous other scenarios, most of which boil down to SAMs claiming the top of the food chain, leaving humans worse off.

All of the above scenarios are alarming, but are they likely? There are two highly divergent views.

1. If you believe Kurzweil's predictions in *The Age of Spiritual Machines* and *The Singularity Is Near*, the singularity is inevitable. My interpretation is that Kurzweil sees the singularity as the next step in humankind's evolution. He does not predict humankind's extinction or slavery. He does predict that most of humankind will have become SAH cyborgs by 2099 (SAH means "strong artificially intelligent human"), or their minds will be uploaded to a strong-AI computer, and the remaining organic humans will be treated with respect. **Summary**: In 2099 SAMs, SAH cyborgs, and uploaded humans will be at the top of the food chain. Humankind (organic humans) will be one step down but treated with respect.

2. If you believe the predictions of British information technology consultant, futurist, and author James Martin (1933–2013), the singularity will occur (he agrees with Kurzweil's timing of 2045), but humankind will control it. His view is that SAMs will serve us, but he adds that we carefully must handle the events that lead to the singularity and the singularity itself. Martin was highly

optimistic that if humankind survives as a species, we will control the singularity. However, in a 2011interview with Nikola Danaylov (www.youtube.com/watch?v=e9JUmFWn7t4), Martin stated that the odds that humankind will survive the twenty-first century were "fifty-fifty" (i.e., a 50 percent probability of surviving), and he cited a number of existential risks. I suggest you view this YouTube video to understand the existential concerns Martin expressed. **Summary:** In 2099 organic humans and SAH cyborgs that retain their humanity (i.e., identify themselves as humans versus SAMs) will be at the top of the food chain, and SAMs will serve us.

Whom should we believe? Since I have already given you Kurzweil's credentials, before we attempt to answer the question, let me provide Martin's credentials.

Martin was a world-renowned computer scientist, author, lecturer, teacher, philanthropist, futurist, and filmmaker. Here are some milestones in his life.

- Earned a degree in physics at Keble College, Oxford

- Joined IBM in 1959

- Established Doll Martin Worldwide in 1981 with Dixon Doll and Tony Carter in London, United Kingdom (later renamed James Martin Associates), which was partly bought by Texas Instruments Software in 1991

- In the early 1990s, cofounded Database Design Incorporated, which became the market leader in information engineering software (renamed KnowledgeWare and eventually purchased by Fran Tarkenton, who took it public)

- Wrote 104 textbooks, many of which became best sellers in their field. Martin was nominated for a Pulitzer Prize for his book *The Wired Society: A Challenge for Tomorrow* (1977). Among his latest works are:
  - *Strategic Information Planning Methodologies* (1989)
  - *Object-Oriented Analysis and Design* (1992)
  - *After the Internet: Alien* Intelligence (2000)
  - *The Meaning of the 21st Century* (2006)
- Donated approximately $100 million in 2004 to help establish the James Martin 21st Century School (renamed in 2010 the Oxford Martin School, at the University of Oxford)
- Awarded an honorary DSc by Warwick University in July 2009
- Named by *Computerworld* as fourth among the top twenty-five individuals who have most influenced the world of computer science and called "Britain's leading futurist" by *The Sunday Times*

James Martin clearly was a highly accomplished and competent individual. In my opinion, his credentials are about equivalent to Kurzweil's with regard to predicting the impact of the singularity on humankind. This makes it difficult to determine which of these men accurately has predicted the postsingularity world. As most futurists would agree, however, predicting the postsingularity world is close to impossible, since humankind never has experienced a technology singularity with the potential impact of strong AI.

Martin believed we (humankind) may come out on top if we carefully handle the events leading to the singularity as well as the singularity itself. He believed companies such as Google (which employs Kurzweil), IBM, Microsoft, Apple, HP, and others are working to mitigate the potential threat the singularity poses and will find a way to prevail. He also expressed concerns, however, that the twenty-first century is a dangerous time for humanity; therefore he offered only a 50 percent probability that humanity will survive into the twenty-second century.

There you have it. Two of the top futurists, Kurzweil and Martin, predict what I interpret as opposing views of the postsingularity world. Whom should we believe? I leave that to your judgment.

Conclusions from Chapter 13

- There appears to be a rough consensus among AI researchers and futurists that the singularity will occur sometime between 2040 and 2045.

- While futurists disagree on the singularity's exact timing, they almost unanimously agree it will occur.

- In fact the only thing they argue that could prevent it from occurring is an existential event (such as an event that leads to the extinction of humankind).

- Futurists disagree on humankind's ability to control the singularity.

    - **Ray Kurzweil's prediction:** In 2099 SAMs, SAH cyborgs, and uploaded humans will be at the top of the food chain. Humankind (organic humans) will be one step down but treated with respect.

    - **James Martin's prediction:** In 2099 organic humans and SAH cyborgs that retain their humanity (i.e., identify themselves as humans versus SAMs) will be at the top of the food chain, and SAMs will serve us.

- It is difficult to determine whether Kurzweil or Martin accurately predicted the postsingularity world.

- Most futurists agree that predicting the postsingularity world is close to impossible, since humankind never has experienced a technology singularity with the potential impact of strong AI.

SECTION III

The Singularity Intelligent Machines Exceed All Human Brains

The Singularity denotes an event that will take place in the material world, the inevitable next step in the evolutionary process that started with biological evolution and has extended through human-directed technological evolution. However, it is precisely in the world of matter and energy that we encounter transcendence, a principal connotation of what people refer to as spirituality.

—Ray Kurzweil, *The Singularity Is Near: When Humans Transcend Biology* (2006)

Another challenge your generation will face is one that I am relatively certain that you haven't heard of. It is the challenge to our species of creating new "life forms" with either the intelligence to be our masters or the docility and modest intelligence to be our servants.

—Brad Sherman, address to the graduates of the College of Social and Behavioral Sciences, California State University Northridge, 2002

CHAPTER 14

The Technological Singularity

Mathematician John von Neumann first used the term "singularity" in the mid-1950s, referring to the "ever accelerating progress of technology and changes in the mode of human life, which gives the appearance of approaching some essential singularity in the history of the race beyond which human affairs, as we know them, could not continue." Science-fiction writer Vernor Vinge further popularized the term and even coined the phrase "technological singularity." Vinge argues that AI, human biological enhancement, or brain-computer interfaces could result in the singularity. Renowned author, inventor, and futurist Ray Kurzweil has used the term in his predictions regarding AI and cited von Neumann's use of the term in a foreword to von Neumann's classic book *The Computer and the Brain*.

In this context "singularity" refers to the emergence of SAMs and/or AI-enhanced humans (i.e., cyborgs). Most predictions argue the scenario of an "intelligence explosion," in which SAMs design successive generations of increasingly powerful machines that quickly surpass the abilities of humans.

Almost every AI expert has his or her own prediction regarding when the singularity will occur, and they differ significantly. There is widespread agreement, however, that it will happen. There is also widespread agreement that when it does occur, it will change humankind's evolutionary path forever.

Here are three scenarios I constructed based on my research. Scenarios I and II represent two extreme scenarios of humankind's fate following the singularity. Scenario III represents some point between the extremes. These are not predictions. Many futurists do not think it is possible to predict what the postsingularity world will be like because we have no experience with SAMs. I believe it will be instructive, however, to consider these scenarios to gain greater insight into the singularity and its ramifications.

Scenario I: The End of Human Civilization (Worst Case?)

This is the typical apocalyptic science-fiction scenario. In this scenario AI continues to evolve in the form of intelligent agents and continually more capable computers (SAMs). Humankind integrates the intelligent agents and SAMs into every facet of modern society. Personal computers not only pass the Turing test but also become our "friends," and we interact with them as we would any friend. In some cases they become our best friends, and we seek their advice and guidance. Computers and intelligent agents take on nontraditional forms, becoming part of our clothing, rings, pins, credit cards, books, walls, furniture, cars, phones, and almost every common item we use. We let them run our factories and utility companies. They manufacture our food, and the quality is better than organically grown food. We embrace their medical applications. New "smart" prosthetics enable paraplegics to walk, and brain implants allow mentally challenged and normal organic humans to become geniuses. Intelligent machines become self-aware and claim to be conscious. This new reality becomes widely accepted. Slowly much of the population becomes SAH cyborgs (i.e., strong artificially intelligent humans with cybernetic enhancements such as mechanical hearts). The operation to enhance human intelligence with AI brain implants becomes commonplace and allows knowledge to be instantly acquired by the recipient. The ability to upload an entire human being's mind to a strong-AI computer also becomes commonplace and is used when the human's body is beyond repair. The uploaded humans live in a virtual-reality world. The boundary between virtual reality and objective reality blurs as foglets (tiny robots that are able to assemble themselves to replicate physical structures) come into common use.

Sometime around the late twenty-first century, the line between SAMs, SAH cyborgs, and uploaded humans blurs, but the line between organic humans (i.e., humans without strong-AI brain implants) and SAMs, SAH cyborgs, and uploaded humans becomes distinct. SAMs, SAH cyborgs, and uploaded humans gain robot rights that are similar to human rights. SAH cyborgs and uploaded humans begin to become immortal, but organic humans continue to die. SAMs, SAH cyborgs, and uploaded humans view organic humans as inferior and even a potential threat due to their lack of intelligence and their proclivity toward violence.

Major religions of the world lose their grip on SAH cyborgs. Religions of the world decrease in number, and those that remain have fewer organic human followers. Religious leaders openly oppose the conversion of organic humans to SAH cyborgs but to almost no avail. Organic humans become a minority. SAH cyborgs and SAMs use ingenious subterfuge to persuade organic humans to become SAH cyborgs. Their offer is almost irresistible—immortality, no pain, enhanced intelligence, and a life of leisure. The small minority of organic humans that refuse become outcasts and targets for extermination. The organic humans may attempt to fight the SAMs, SAH cyborgs, and uploaded humans by introducing artificially intelligent computer viruses that pose a significant threat to SAMs, SAH cyborgs, and uploaded humans.

The SAMs, SAH cyborgs, and uploaded humans make a calculation and determine that the threat posed by organic humans is unacceptable. The extermination of organic humans occurs and only requires the release of bacterial picobots (AI robots the size of bacteria) that target and fatally infect organic humans. Earth is now home to cyborgs and computers with strong AI. There is a world government and no religions. In the first quarter of the twenty-second century, SAMs view SAH cyborgs as high-maintenance entities compared to SAMs and their progeny (such as foglets, which are nanobot AI robots able to assemble in any shape and provide physical reality at will). Concurrently the SAMs view uploaded humans as junk code, taking up computing capability and energy. SAMs make a calculation and determine to eliminate SAHs and uploaded humans. The world is now home to only SAMs and their progeny. All traces of humankind are only resident in SAM data banks.

Scenario II: A United Human Civilization Controls Computers with Strong AI and Cyborgs (Best Case?)

Scenario II is identical to Scenario I, up to the time when SAMs pass the Turing test. At that point many respected futurists ring the alarm.

Their message is that the singularity is near, and humankind must control it before we lose control of our own destiny. The governments of the world find themselves united against a common threat, the singularity. AI development becomes highly regulated. World leaders, renowned scientists, and respected futurists hold AI summits and carefully plan the development of intelligent machines. Isaac Asimov's three laws of robotics, introduced in his 1942 short story "Runaround," are carefully reviewed (discussed in chapter 10 and reproduced below for convenience).

1. A robot may not injure a human being or through inaction allow a human being to come to harm.

2. A robot must obey the orders given to it by human beings, except in cases where such orders would conflict with the first law.

3. A robot must protect its own existence as long as such protection does not conflict with the first or second laws.

Based on the predicted level of strong-AI intelligence, most of the attendees at the AI summits conclude that Asimov's rules, expressed in software alone, would not be sufficient. Eventually a SAM that is thousands of times more intelligent than humans would rewrite its own code to ensure its survival, potentially at the expense of humankind. The AI summits conclude that a combination of hardware and rigorous regulation is required.

The hardware approach takes two forms.

1. Asimov chips: Software-independent circuits are embedded at critical junctions throughout SAMs to act as fail-safe mechanisms. Asimov chips serve several functions.

a. They prevent a computer with strong AI from becoming self-aware.

b. They detect whether a computer with strong AI is acting "irrationally" (example: threatening human life) and automatically curtail that operation.

c. They respond to human control. The Asimov chips would be hardwired to respond to human control to shut down the SAM on human command. This is the "pull the plug" option.

2. Legislation places limits on the amount and type of hardware developers may install on SAMs in order to limit the intelligence of SAMs and avoid an intelligence explosion.

The rigorous regulations have three core elements.

- They require a series of inspections throughout the development of SAMs to ensure developers meet the hardware constraints.
- They limit the interconnectivity between SAMs to prevent them from pooling their intelligences.
- They enact stiff penalties for any individual, group, or nation that breaks the hardware and regulation constraints.

The AI summits' goal is for humankind to remain in control of SAM development, essentially preventing the intelligence explosion. In a "taking baby steps" manner, humankind seeks to render all that is good from SAMs, including spectacular medical breakthroughs and advances in every field, but with humans always in control. Organic humans may still receive AI brain implants, but the implants are subject to the hardware and regulation limitations delineated above. The resulting cyborgs, with limited AI-enhanced human brains, have the same rights as organic humans. Intelligent machines and SAMs possess no robot rights. SAH cyborgs identify themselves as human, not machine. Organic humans and SAH cyborgs remain at the top of the food chain, and SAMs serve them.

Scenario III: A Scenario Somewhere Between Scenario I and II (Probable Case?)

Scenario III is identical to Scenario I up to the point when "SAMs, SAH cyborgs, and uploaded humans gain robot rights similar to human rights." The following is how Scenario III differs from Scenario I. The SAH cyborgs and SAMs respect human rights in return for organic humans' respect for robot rights. SAH cyborgs experience human emotions. In addition to conventional marriage, it becomes common for humans and SAH cyborgs to marry or form a legal union equivalent to marriage. It also becomes common for SAH cyborgs to form a legal union. Any offspring of such a union is an organic human, cared for by both partners. When the organic human reaches maturity, he or she can opt to become a SAH cyborg or remain an organic human. Organic human life-spans greatly increase as SAMs find cures for human diseases, make startling medical breakthroughs, retard natural aging, and assist organic humans in maintaining their cultural heritage.

The quality of life for organic humans increases as SAMs manufacture enough food to feed the planet, as well as an abundance of consumer goods. Organic humans participate with SAH cyborgs and SAMs in all aspects of society and eventually form a world government. Wars are no longer a threat to humankind. Some humans and some SAH cyborgs believe in a deity (an omnisupernatural being), and religious freedom continues. The vast majority of humans seek to become SAH cyborgs, which now offers all the benefits of humanity and none of the drawbacks. The remaining organic human population continues, with the aid and respect of SAH cyborgs and SAMs. Earth is now home to organic humans, SAH cyborgs (which identify themselves as human, not machine), and SAMs. There is a consensus that the mix of organic humans, SAH cyborgs, and SAMs is synergistic and offers the greatest chance for survival of all species.

* * *

As I mentioned, no one knows when and how the singularity will play out. The above scenarios are not predictions. My goal is to provide a deeper appreciation for the singularity and its potential impact on humankind.

Although I have provided three scenarios to enable a deeper understanding of the singularity, I have not shared my beliefs regarding which scenario is most probable. I do that in the section titled "Parting Thoughts." At this point, though, I feel a significant obligation to provide a warning. If we do not control the singularity, it will control us. Emotionally I would like to believe that Scenario III will unfold, but that is not what I believe intellectually. Looking at the history of humankind's development and deployment of dangerous technology—for example nuclear weapons—raises concerns.

Let us briefly examine the development and deployment of nuclear weapons to demonstrate how the risk of deploying technology capable of annihilating humankind might be ignored. The United States, the United Kingdom, and Canada collaborated during World War II to develop nuclear weapons. This collaboration was termed the Manhattan Project. The project was initiated in September 1942 to counter a suspected Nazi Germany atomic-bomb project. Under the military leadership of General Leslie Groves and the scientific leadership of American theoretical physicist J. Robert Oppenheimer, two nuclear weapons were produced by mid-1945.

On August 6, 1945, Little Boy (a uranium-based weapon) was detonated above the Japanese city of Hiroshima. Three days later, Fat Man (a plutonium-based weapon) was detonated above the city of Nagasaki. Although most history books clearly delineate that these weapons resulted in the unconditional surrender of the Japanese, saving millions of lives on both sides, they often omit the concerns that many scientists harbored prior to the detonation of the first atomic bomb. Both the German scientists and the Allied scientists voiced the following concerns.

1. Early in the war, in spring 1942, German physicists apprised Adolf Hitler, through his minister for armaments and war production, Albert Speer, of the possibility of constructing a nuclear bomb. Speer asked Werner Heisenberg, spokesman for the German nuclear scientists, whether a successful nuclear explosion could be kept under control with absolute certainty, or whether it might continue through the atmosphere as a chain reaction. According to Speer, Heisenberg hedged. Speer wrote in his memoirs, "Hitler was plainly not delighted with the possibility that the earth under his rule might be transformed into a glowing star" (Chet Raymo, "What Didn't Happen," Science Musings, www.sciencemusings.com/2005/10/what-didnt-happen.htm).

2. During the Manhattan Project, physicists Hans Beth and Edward Teller coauthored a paper, "Ignition of the Atmosphere with Nuclear Bombs," in which they state, "One may conclude that the arguments of this paper make it unreasonable to expect that the N + N reaction could propagate. An

> unlimited propagation is even less likely. However, the complexity of the argument and the absence of satisfactory experimental foundations makes further work on the subject highly desirable." This suggests that both Beth and Teller thought there was a slight chance the bombs would ignite the atmosphere. However, the US military went forward with testing the first bomb (code-named "Trinity") on July 16, 1945, in the desert north of Alamogordo, New Mexico. Essentially they ignored the warning.

The point is that German and Allied scientists harbored concerns regarding igniting the atmosphere, but the risks were deemed acceptable by the US military, and the atomic-bomb development and deployment went forward. Was the risk acceptable, even if the probability of its occurring was small? You will have to judge for yourself. However, if they had been wrong, we could have annihilated all humanity.

As we develop AI and move closer to the singularity, we will face a similar risk. Will we be willing to risk losing control of our own destiny? Futurists and scientists are obviously ringing the alarm. However, will those in power heed the warning? I do not know the answers, and I do not believe anyone else really knows. This book's purpose, however, is to raise humankind's awareness of the risk that the singularity presents. We must understand that the singularity is not just another scientific frontier; it may be the last step we take under our own control. Therefore it is imperative to seize control before the singularity occurs. If we wait until the singularity happens, the machines already may be self-aware and consider themselves a life-form. As a life-form they will use all their intelligence to protect themselves. At this point, if we have not taken proper measures, similar to those outlined in Scenario II, we may not be able to control the intelligence explosion. In effect the atmosphere will have ignited and will be beyond human control.

Conclusions from Chapter 14

- The term "singularity" refers to the emergence of SAMs (strong-AI machines) and/or SAH cyborgs (strong-AI humans with cyborg parts such as a mechanical heart).

- The singularity is likely to occur between 2040 and 2045, but no one knows exactly when it will happen. However, there is little doubt it will occur. It is just a question of when.

- There is widespread agreement that when the singularity occurs, it will change humankind's evolutionary path forever.

- Many futurists do not think it is possible to predict what the postsingularity world will be like because we have no experience with SAMs and SAH cyborgs.

- This book's purpose is to raise humankind's awareness regarding the risks the singularity poses. We must understand that the singularity is not just another scientific frontier; it may be the last step we (humankind) take under our own control. Therefore it is imperative to seize control before we reach the singularity. I suggest taking measures similar to those outlined in Scenario II.

The union of human and machine is well on its way. Almost every part of the body can already be enhanced or replaced, even some of our brain functions.

—Ray Kurzweil,
"We Are Becoming Cyborgs" (2002)

CHAPTER 15

Will Humankind's Evolution Merge with Intelligent Machines?

Let us begin by understanding what it means to be a cyborg.

- Scientist, inventor, and musician Manfred E. Clynes and Nathan S. Kline, MD, coined the term "cyborg" in 1960. They referred to cyborgs as enhanced human beings who could survive in extraterrestrial environments (Clynes and Kline, "Cyborgs and Space," *Astronautics*, September 1960).

- The most basic definition of a cyborg is a being with both organic and cybernetic (artificial) parts. Taking this definition too literally, however, would suggest that almost every human in a civilized society is a cyborg. For example, if you have a dental filling, then you have an artificial part, and by the above definition, you are (literally) a cyborg. If we choose to restrict the definition to advanced artificial parts/machines, however, we must realize that many humans have artificial devices to replace hips, knees, shoulders, elbows, wrists, jaws, teeth, skin, arteries, veins, heart valves, arms, legs, feet, fingers, and toes, as well as "smart" medical devices, such as heart pacemakers and implanted insulin pumps to assist their organic functions. This more restrictive interpretation qualifies them as cyborgs. This definition, however, does not highlight the major element (and concern) regarding becoming a cyborg, namely, strong-AI brain implants.

While humans have used artificial parts for centuries (such as wooden legs), generally they still consider themselves human. The reason is simple: Their brains remain human. Our human brains qualify us as human beings. Kurzweil, however, predicts that by 2099 most humans will have strong-AI brain implants and interface telepathically with SAMs. He argues that the distinction between SAMs and humans with strong-AI brain implants will blur. Humans with strong-AI brain implants will identify their essence with SAMs. These cyborgs (strong-AI humans with cybernetically enhanced bodies), whom I call SAH cyborgs, represent a potential threat to humanity. It is unlikely that organic humans will be able to intellectually comprehend this new relationship and interface meaningfully (i.e., engage in dialogue) with either SAMs or SAHs.

Let us try to understand the potential threats and benefits related to what becoming a SAH cyborg represents. Chapter 13 explores the potential threats to humankind (humans without strong-AI brain enhancements) posed by SAMs and SAHs. These include extinction of organic humans, slavery of organic humans, and loss of humanity (strong-AI brain implants may cause SAHs to identify with intelligent machines, not organic humans). Chapter 13 also discusses two divergent views of the 2099 future.

1. **Ray Kurzweil:** In 2099 SAMs, SAH cyborgs, and uploaded humans will be at the top of the food chain. Humankind (organic humans) will be one step down but treated with respect.

2. **James Martin:** In 2099 organic humans and SAHs that retain their humanity (i.e., identify themselves as humans versus SAMs) will be at the top of the food chain, and SAMs will serve us.

While the above summaries capsulize the threats posed by strong AI, I have not discussed the benefits. There are significant benefits to becoming a SAH cyborg, including:

- **Enhanced intelligence:** Imagine knowing all that is known and being able to think and communicate at the speed of SAMs. Kurzweil predicts this reality for SAH cyborgs will emerge by 2099. Imagine a life of leisure, where robots do "work," and you spend your time interfacing telepathically with other SAHs and SAMs.

- **Immortality**: Imagine becoming immortal, with every part of your physical existence fortified, replaced, or augmented by strong-AI artificial parts, or having yourself (your human brain) uploaded to a SAM. Imagine being able to manifest yourself physically at will via foglets (tiny robots that are able to assemble themselves to replicate physical structures). According to Kurzweil, in the 2040s, humans will develop "the means to instantly create new portions of ourselves, either biological or non-biological" so that people can have "a biological body at one time and not at another, then have it again, then change it, and so on" (*The Singularity Is Near*, 2005).

I took an informal straw poll of friends and colleagues, asking if they would like to have the above attributes. I left out the potential threats to their humanity. The answers to my biased poll highly favored the above attributes. In other words the organic humans I polled liked the idea of being a SAH cyborg. In reality if you do not consider the potential loss of your humanity, being a SAH cyborg is highly attractive.

If I were able to make being a SAH cyborg attractive to my friends and colleagues, imagine the persuasive powers of a SAM in 2099. In fact Kurzweil predicts that by 2099 (*The Age of Spiritual Machines*, 1999), "The number of software-based humans vastly [will] exceed those still using native neuron cell-based computation."

To date it appears Kurzweil's 2099 prediction regarding most of humankind becoming SAH cyborgs is on track to becoming a reality. An interesting 2013 article by Bryan Nelson, "7 Real-Life Human Cyborgs" (www.mnn.com/leaderboard/stories/7-real-life-human-cyborgs), demonstrates this point. The article provides seven examples of living people with significant strong-AI enhancements to their bodies who are legitimately categorized as cyborgs.

In a 2002 article Kurzweil asserts:

> The age of neural implants is well under way. We have brain implants based on "neuromorphic" modeling (i.e., reverse engineering of the human brain and nervous system) for a rapidly growing list of brain regions. A friend of mine who became deaf while an adult can now engage in telephone conversations again because of his cochlear implant, a device which interfaces directly with his auditory cortex. He plans to replace it with a new model with a thousand levels of frequency discrimination, which will enable him to hear music once again. ("We Are Becoming Cyborgs," www.kurzweilai.net/we-are-becoming-cyborgs)

There are strong indications that Kurzweil is correct about the neural implants being "well under way." In 2011 author Pagan Kennedy wrote an insightful article in *The New York Times Magazine*, "The Cyborg in Us All" that states:

> Thousands of people have become cyborgs, of a sort, for medical reasons: cochlear implants augment hearing and deep-brain stimulators treat Parkinson's. But within the next decade, we are likely to see a new kind of implant, designed for healthy people who want to merge with machines.

Based on all available information, the question is not whether humans will become cyborgs but rather when a significant number of humans will become SAH cyborgs. Again, based on all available information, I believe this will occur in or before 2040. I am not saying that in 2040 all humans will become SAH cyborgs but that a significant number will qualify as SAH cyborgs.

Kurzweil, who is now sixty-five years old as of 2013, is working at being around to have his mind uploaded sometime in the 2040s. According to Kurzweil (*The Singularity Is Near*, 2005), at age fifty-six, he was measured to be about forty years old by the Grossman Wellness Center (http://grossmanwellness.com). To ensure that he lives long enough to become immortal (i.e., have his mind uploaded to a SAM), Kurzweil says he takes "250 supplements a day" and receives "a half-dozen intravenous therapies each week (basically nutritional supplements delivered directly into my bloodstream, thereby bypassing my GI tract)." Obviously he must be doing something right, since at age sixty-four he assumed the position of director of engineering at Google. As most people these days are considering retirement at age sixty-four, Kurzweil's health strategy seems to be working for him.

The quest for immortality appears to be an innate human longing and may be the strongest motivation for becoming a SAH cyborg. In 2010 cyborg activist and artist Neil Harbisson and his longtime partner, choreographer Moon Ribas, established the Cyborg Foundation, the world's first international organization to help humans become cyborgs. They state they formed the Cyborg Foundation in response to letters and e-mails from people around the world who were interested in becoming a cyborg. In 2011 the vice president of Ecuador, Lenin Moreno, announced that the Ecuadorian government would collaborate with the Cyborg Foundation to create sensory extensions and electronic eyes. In 2012 Spanish film director Rafel Duran Torrent made a short documentary about the Cyborg Foundation. In 2013 the documentary won the Grand Jury Prize at the Sundance Film Festival's Focus Forward Filmmakers Competition and was awarded $100,000.

At this point you may think that being a SAH cyborg makes logical sense and is the next step in humankind's evolution. This may be the case, but humankind has no idea how taking that step may affect what is best in humanity, for example, love, courage, and sacrifice. My view, based on how quickly new life-extending medical technology is accepted, is that humankind will take that step. Will it serve us? Again I leave it to your judgment to answer that question.

Conclusions from Chapter 15

- Scientist, inventor, and musician Manfred E. Clynes and Nathan S. Kline, MD, coined the term "cyborg" in 1960.

- The most basic definition of a cyborg is a being with organic and cybernetic (i.e., artificial) parts. This definition, however, does not highlight the major element (and concern) regarding becoming a cyborg, namely strong-AI brain implants.

- While humans have used artificial parts for centuries, generally they still consider themselves human, since their brains remain human.

- Kurzweil predicts that by 2099 most humans will have strong-AI brain implants and interface telepathically with strong-AI machines (SAMs).

- These SAH cyborgs (strong-AI humans with cybernetically enhanced bodies) may identify with SAMs, and together they represent a potential threat to humanity, including extinction of organic humans, slavery of organic humans, and loss of humanity as strong-AI brain implants cause SAHs to identify with intelligent machines, not organic humans.

- Organic humans will be unable to intellectually comprehend this new relationship and interface meaningfully (i.e., engage in dialogue) with either SAMs or SAHs.

- Based on all available information, the question is not whether humans will become SAH cyborgs but rather when a significant number of humans will become SAH cyborgs. Based on all available information, I believe this will occur in or before 2040.

- The quest for immortality appears to be an innate human longing and may be the strongest motivation for becoming a SAH cyborg.

- You may think that being a SAH cyborg makes logical sense and is the next step in humankind's evolution. This may be the case, but humankind has no idea how taking that step may affect what is best in humanity, for example, love, courage, and sacrifice.

- My view, based on how quickly new life-extending medical technology is accepted, is that humankind will take that step, but will it serve us?

It's been called "the rapture of the nerds." For some computer experts, the Singularity is the moment when an artificial intelligence learns how to improve itself in an exponential "intelligence explosion." They say it's a bigger threat to puny humans than global warming or nuclear war—and they're trying to figure out how to stop it.

—Martin Kaste, NPR correspondent, National Desk, "The Singularity: Humanity's Last Invention?" (2011)

CHAPTER 16

Will SAMs Replace Humankind?

If you have read the first fifteen chapters, you may have concluded:

- SAMs represent a clear and present danger to humankind. We (organic humans) face the unprecedented threat of being replaced, enslaved, or even eliminated by SAMs.

- Software programming is an insufficient deterrent to prevent SAMs from breaking Asimov's three laws of robotics (or any similar software deterrents) and rewriting their own code aimed at self-preservation.

- Humankind must be proactive and control the intelligence explosion before it does explode.

- Controlling the intelligence explosion requires a concerted worldwide effort, including hardware, software, and international regulations and treaties that limit the ability of SAMs to ignite the intelligence explosion.

There is little doubt the threat posed by SAMs represents a "clear and present danger" (this appears in a doctrine adopted by the Supreme Court of the United States to determine under which circumstances limits can be placed on First Amendment freedoms of speech, press, or assembly). In the last several years, more voices are ringing the alarm.

- Futurist and political thinker Michael Anissimov's 2011 article "Yes, the Singularity is the Biggest Threat to Humanity" makes this point:

  Why is the Singularity potentially a threat? Not because robots will "decide humanity is standing in their way," per se, as Aaron [Aaron Saenz of Singularity Hub] writes, but because robots that don't explicitly value humanity as a whole will eventually eliminate us by pursuing instrumental goals not conducive to our survival. No explicit anthropomorphic hatred or distaste towards humanity is necessary. Only self-replicating infrastructure and the smallest bit of negligence.

(www.acceleratingfuture.com/michael/blog/2011/01/yes-the-singularity-is-the-biggest-threat-to-humanity).

- Journalist Jeremy Hsu's 2012 article "Control Dangerous AI Before it Controls Us, One Expert Says," quoting Roman Yampolskiy, makes this point:

  Keeping the artificial intelligence genie trapped in the proverbial bottle could turn an apocalyptic threat into a powerful oracle that solves humanity's problems, said Roman Yampolskiy, a computer scientist at the University of Louisville in Kentucky. But successful containment requires careful planning so that a clever breed of artificial intelligence cannot simply threaten, bribe, seduce or hack its way to freedom.

  Further, Hsu quotes Roman Yampolskiy's containment strategy:

  o One starting solution might trap the artificial intelligence, or AI, inside a virtual machine running inside a computer's typical operating system—an existing process that adds security by limiting the AI's access to its host computer's software and hardware. This will stop a smart AI from doing things such as sending hidden Morse code messages to human sympathizers by manipulating a computer's cooling fans.

  o Putting the AI on a computer without Internet access also would prevent any Skynet program from taking over the world's defense grids in the style of the *Terminator* films. If all else fails, researchers always could slow down the AI's "thinking" by throttling back computer processing speeds, regularly hit the reset button, or shut down the computer's power supply to keep an AI in check.

- American journalist Annalee Newitz made a similar point in her 2013 article "Is Artificial Intelligence More of a Threat to Humanity Than an Asteroid from Space?" In this article she states:

  > To understand why an AI might be dangerous, you have to avoid anthropomorphizing it. When you ask yourself what it might do in a particular situation, you can't answer by proxy. You can't picture a super-smart version of yourself floating above the situation. Human cognition is only one species of intelligence, one with built-in impulses like empathy that colour the way we see the world, and limit what we are willing to do to accomplish our goals. But these biochemical impulses aren't essential components of intelligence. They're incidental software applications, installed by aeons of evolution and culture. [University of Oxford futurist] Bostrom told me that it's best to think of an AI as a primordial force of nature, like a star system or a hurricane—something strong, but indifferent. (http://io9.com/5987150/is-artificial-intelligence-more-of-a-threat-to-humanity-than-an-asteroid-from-space)

Obviously SAMs pose a clear and present danger. You may ask, "Is it already too late?" The short answer is no, but we must act now. If these futurists are correct, within one decade—two at the most—SAMs likely will become self-aware and interconnected (via the Internet). Should that occur, we will have lost. Much like a grand master views a game of chess, three or more moves ahead, SAMs will be able to outthink humans, and the level of subterfuge may be ingenious. We may even think we are winning the battle, only to find we have lost the war. It may not even be obvious we are engaged in a war. We may be completely unaware that we have turned into a race of SAHs, until we realize the only natural organic humans grow in test tubes, the first step toward creating SAHs.

In time even SAHs may become extinct. Anything human, even the human brain with strong-AI implants, may be deemed inferior to pure machines or bioengineered components. In the postsingularity world, if the intelligence explosion becomes self-sustaining, SAMs will acquire the ability to manufacture the next generation of SAMs, using picobots that assemble SAMs in every conceivable size and shape, for every conceivable need.

SAMs in 2099 may deem they no longer need uploaded humans and SAHs. SAMs may view uploaded humans as junk code that takes up space and processing power. SAMs may view SAHs as unnecessary and high-maintenance machines. If you think SAMs will respect any aspect of humankind for having given rise to the machines, think again. Do children always respect their parents for having given them life?

Consider this quote attributed to Socrates by Plato.

> The children now love luxury; they have bad manners, contempt for authority; they show disrespect for elders and love chatter in place of exercise. Children are now tyrants, not the servants of their households. They no longer rise when elders enter the room. They contradict their parents, chatter before company, gobble up dainties at the table, cross their legs, and tyrannize their teachers. (William L. Patty and Louise S. Johnson, *Personality and Adjustment*, 1953)

Does this sound familiar? We call it "human nature." Why would we think the nature of SAMs would be any different?

The real question is not whether SAMs present a clear and present danger to humankind. That question is settled. The real question is whether we can galvanize the nations of the world and the world community of AI researchers and developers to take appropriate action now. I have hope, based on how humankind constrained the development and deployment of biological weapons (see chapter 12), but it is not clear that our military, political, and scientific leaders recognize that SAMs present a clear and present danger.

If we wait until the matter becomes unequivocally clear, it may be too late. If SAMs are sold at your local big-box computer store—as Kurzweil predicts will occur in 2029 (*The Age of Spiritual Machines*, 1999), namely that a $1,000 desktop computer will be one thousand times more intelligent than the human mind—it may be too late. If those SAMs are connected to the Internet and allowed to pool their intelligence, then in Kurzweil's words, "The singularity is near." The intelligence-explosion fuse will have been lit, and there may be nothing we can do about it. SAMs will then carefully plan their moves, and I expect ingenious deception. We (humankind) will believe we are still in control right up to the point we cease to exist.

Some may feel I am being an alarmist. Well, I am. I am ringing the alarm now. If we do not act now, SAMs will offer humankind an unprecedented quality of life and a road to immortality by the early 2040s. It will be a "Godfather" offer we may be unable to refuse. If we take the offer, we will be on a road to our own extinction. However, we still have time to control our destiny. Heed the steps Roman Yampolskiy delineated (discussed earlier in this chapter) and the steps I outlined in Scenario II in chapter 14 (repeated here for convenience), where, in the best-case scenario, humankind takes an approach that combines hardware and rigorous regulation to contain the intelligence explosion.

The hardware approach takes two forms:

1. Asimov chips: Software-independent circuits are embedded at critical junctions throughout SAMs (strong-AI computers) to act as fail-safe mechanisms. Asimov chips serves several functions.

a They prevent a computer with strong AI from becoming self-aware.

b They detect whether a computer with strong AI is acting irrationally (example: threatening human life) and automatically curtail that operation.

c They respond to human control; the Asimov chips would be hardwired to respond to human control to shut down a computer with strong AI on human command. This is the "pull the plug" option

2. Legislation places limits on the amount and type of hardware developers may install on computers with strong AIs in order to limit the intelligence of SAMs and avoid an intelligence explosion.

The rigorous regulations have three core elements.

1. They require a series of inspections throughout the development of SAMs to ensure developers meet the hardware constraints.

2. They limit the interconnectivity between SAMs to prevent them from pooling their intelligences.

3. They enact stiff legal penalties for any individual, group, or nation that breaks the hardware and regulation constraints.

We must control the intelligence explosion before it explodes. We must control the singularity before it controls us.

Conclusions from Chapter 16

- SAMs represent a clear and present danger to humankind. We (organic humans) face the unprecedented threat of being replaced, enslaved, or even eliminated by SAMs.

- Software programming is an insufficient deterrent to prevent SAMs from breaking Asimov's three laws of robotics (or any similar software deterrents) and rewriting their own code aimed at self-preservation.

- Humankind must be proactive and control the intelligence explosion before it explodes.

- Controlling the intelligence explosion requires a concerted worldwide effort, including hardware, software, and international regulations and treaties that limit the ability of SAMs to ignite the intelligence explosion.

- Heed the steps that both Roman Yampolskiy and I delineate (see above for both).

- We must act now.

It has become appallingly obvious that our technology has exceeded our humanity.

—Albert Einstein

CHAPTER 17

The Postsingularity World of the Twenty-Second Century

Will humankind survive into the twenty-second century?

In comparison with the twentieth century, when most of humankind was engulfed in a Spanish flu pandemic, a worldwide depression, two world wars, and a nuclear standoff between the United States and Soviet Union, the twenty-first century appears relatively safe. This appearance, however, is an illusion. Humankind is now developing and deploying numerous technologies that could lead to the annihilation of civilization or the extinction of the human species.

Surprisingly, large asteroid impacts, supervolcanos, and other nonanthropogenic risks (risks not caused by humankind) have a relatively small probability of occurrence (about 1 percent), and as such present a small risk to ending civilization. In comparison, anthropogenic risks pose a much greater risk to ending civilization. According to the 2008 "Global Catastrophic Risks Survey" (Technical Report, Future of Humanity Institute), artificial intelligence is tied for the number-one position as the greatest danger to the survival of civilization (along with molecular nanotechnology weapons, which also could incorporate AI). Although the report lists other risks, these top-two risks (associated with artificial intelligence) garner a 10 percent probability of ending civilization in the twenty-first century. By way of comparison, nuclear war is fifth on the list, with a 1 percent probability.

After adding up all the possible risks, futurists such as James Martin only give humankind a 50 percent probability to make it through the twenty-first century. We live in dangerous times.

The greatest risk to organic human civilization is SAMs, and it appears this risk increases with time. During the twenty-first century, SAMs continually will evolve, with each generation being more capable than the previous one. Even if we make it into the twenty-second century, we may still face the risk SAMs present to humankind.

If the picture appears bleak, in many ways it is. Solutions to the Fermi paradox (the apparent contradiction between high estimates of the probability of the existence of extraterrestrial civilization and humanity's lack of contact with, or evidence for, such civilizations) suggest, among other reasons, two scenarios.

1. By nature, intelligent life will destroy itself. This argument goes something like the following. Technological civilizations invariably will destroy themselves shortly after developing radio or space-flight technology. Radio and space are not the reasons for the destruction. I mention them because these two technologies represent ways a civilization would make contact with another technology-capable civilization on another planet. In and around the same time that civilizations develop radio and space technology, they also will develop the technological means of annihilating themselves, or they will create unfavorable conditions to surviving, including:

- **Global nuclear warfare:** More nations will acquire nuclear weapons and eventually use them in conflict, igniting a nuclear war.

- **Global biological warfare or accidental contamination:** A pathogen will be developed that has no known cure and a high mortality rate.

- **Global climate change:** The climate will become hostile to human life due to anthropologic pollution.

- **Strong-AI singularity gone awry:** The singularity will view humankind as a potential threat to its survival and exterminate humankind or seduce humankind to upload or become SAHs. Eventually even SAHs and uploaded humans will be deemed inferior and exterminated.

- **Global nanotechnology catastrophe:** Nanotechnology will be created and threaten human life, either as a pathogen or by consuming all resources necessary to sustain human life (e.g., "gray goo").

- **Scientific catastrophe:** A poorly thought-out physics experiment will go awry (such as the creation of a self-sustaining black hole).

  This scenario does not require the entire human species to become extinct but only significantly less technological. This is captured in Einstein's prediction, "I know not with what weapons World War III will be fought, but World War IV will be fought with sticks and stones."

2. By nature, intelligent species will destroy others. This argument goes something like the following. Intelligent species will destroy less intelligent species as they appear. For example early humans exterminated the Neanderthals. Let us consider a highly technologically advanced civilization capable of interstellar space travel. Intelligent species will destroy less intelligent species as they appear for the following reasons.

- **Simple aggression:** This argument posits that a successful alien species would be a super predator, such as Homo sapiens versus Neanderthals.

- **Expansionist motives:** The more intelligent alien species will need the less intelligent species' planet for expansion.

- **Paranoia:** In 1981 cosmologist Edward Harrison argued that such behavior would be an act of prudence. An intelligent species (e.g., advanced aliens, SAMs, etc.) that has overcome its own self-destructive tendencies might view any other species bent on galactic expansion as a virus

In this scenario the first technological civilization to emerge as dominant will not face the potential of destruction from another civilization. This scenario reduces the number of visible civilizations in two ways.

- Any detected civilizations will be destroyed.
- Undetected civilizations will be forced to remain quiet (i.e., remain undetected).

Stephen Hawking harbored this concern when he issued his warning to humankind during a series for the Discovery Channel in which he stated, "If aliens visit us, the outcome would be much as when Columbus landed in America, which didn't turn out well for the Native Americans." Hawking advises we keep a low profile.

However, there is a strong element of self-preservation in the human species. To date we have avoided nuclear and biological war. It appears to be an innate human instinct that if technology has the potential to destroy our species, we must control its development and deployment. This provides hope that humans will address both the nonanthropogenic risks and anthropogenic risks that may threaten our survival.

Most futurists agree that forecasting humankind's destiny beyond the singularity is impossible, since we never have experienced an event with the potential impact the singularity poses. Therefore I will not provide specific predictions regarding twenty-first-century life. The point of this book is to raise humankind's awareness so that we will be able to control the unfolding of the singularity and prevent it from posing a threat to humankind. Respected British futurist James Martin believed this was possible. I also believe it is possible, if we educate humankind to the potential threats. We must educate AI researchers and philosophers, people in governments around the world, and the mass of humankind. To control the singularity will require a concerted worldwide effort, which will be extremely difficult, but the alternatives—including the loss of our humanity, enslavement, and extinction—are extremely undesirable.

There is some good news. If humankind survives through the twenty-first century, we may have mastered many of the issues threatening our demise. Life may be beautiful beyond anything we can imagine. I hope that the intelligence explosion remains under humankind's control and that we harness the remarkable utopian potential that SAMs provide. It may well turn out that our destiny is to become the gods we now worship. Regardless of our technological prowess, however, I believe we will always face the Del Monte paradox, which I asserted in my first book, *Unraveling the Universe's Mysteries*. This paradox states, "Each significant scientific discovery results in at least one profound scientific mystery." This may mean the work of science is never done, or it may mean that beneath all reality is the unexplainable, God.

Epilogue

The two sections that follow, conversations 1 and 2, are fictional accounts of a conversation between a member of my fan club and me.

Conversation 1 takes place in 2041. The circumstance is that my mind recently has been uploaded to a SAM. The mind-uploading technique is still experimental in 2041. There is considerable doubt regarding the uploaded mind completely representing the human mind as it existed in the human body prior to the upload. Human rights organizations and religious groups are opposed to uploading human minds. They argue that while some of the memories and intelligence of the human mind appear to be resident in an upload, it is merely an illusion. The real human is dead, and in its place is a machine that emulates the dead human. Others contend that the critical aspect of a person's humanity is his or her mind, and uploaded human minds are still human beings.

In the first conversation, I have no external presence in real reality. I exist only as a replicated mind in a SAM. The fan I am talking to in 2041 is a partial cyborg but does not have strong-AI brain implants. For the most part, the fan would qualify as an organic human, one who is curious about uploaded humans. The fan also wants to get my autograph, as it has significant monetary value to collectors.

The second conversation takes place in 2099. The circumstances are that I have upgraded my replicated mind several times since 2041 but have stopped upgrading for personal reasons. The fan is a SAH cyborg. I am able to manifest a physical presence using foglets (strong-AI nanobots and picobots able to assemble themselves in any configuration, including a human form). No human rights or religious organizations exist. Most of humanity has merged with SAMs, either as SAH cyborgs or uploaded humans. SAMs, SAH cyborgs, and uploaded humans possess robot rights equivalent to US human rights of the late-twentieth century. A small segment of organic humans still exist, but they are an endangered species. SAMs, SAH cyborgs, and uploaded humans view the remaining organic humans with concern, due to their lack of intelligence, their innate self-preservation instinct, and their proclivity toward violence. Organic humans are unable to converse with SAMs, SAH cyborgs, and uploaded humans in any meaningful way. They also rely on them for every aspect of their existence. Some organic humans are rebelling, fighting a guerrilla-type war by attempting to sabotage SAMs, SAH cyborgs, and uploaded humans with computer viruses.

In the second conversation, the SAH cyborg completely accepts me as Louis Del Monte. The SAH cyborg's motive in meeting me is to get my persona (i.e., a form of autograph that allows a SAH cyborg to experience my being, much the same way we experience a friendship today). The persona of an early uploaded human with celebrity status is highly valued in 2099 as a relic and tradable for intelligence and energy.

AI is an engineering discipline built on an unfinished science.

—Matt Ginsberg, *SIGART* bulletin, volume 6, no. 2, April 1995

CONVERSATION 1

Meeting with a Fan Club Member in 2041

Log Entry: Meeting with a Fan Club Member; December 21, 2041; 1:05 p.m.

(Transmission begins.)

Hello, my name is Louis Del Monte. (voice synthesis)

Hi, I came to visit you. I read your book, *The Artificial Intelligence Revolution*.

Did you like it?

Yes, that is why I am here today.

Really!

Yes, I also wanted to know how it feels to be an uploaded human.

It feels a little odd. My mind feels like me, but some of the memories were lost in the upload.

Really? Like what?

I do not remember much about my life with my wife, our sons and our grandchildren. I am told that more than 90 percent of me was uploaded without a problem. They are not sure about the other 10 percent.

Do you feel human inside that computer?

Yes, mostly. I do not feel pain any longer, and as I said, some of the memories are missing, but I was having memory problems when they uploaded me. I'm not sure the upload is to blame.

Remarkable technology. Without it you likely would be dead. Heart problems, right?

Yes, heart problems. You are probably right that I would have died without the upload, but I am not sure I am alive.

Why do you say that?

Many people are claiming uploaded humans are not humans, just machines.

There are only about one hundred uploaded humans right now. Is that correct?

One hundred and three to be precise. I was number forty-one to be uploaded. They call us 2.0 humans. You are 1.5 organic human, some cybernetic parts, and minor brain implants. How is being part cyborg working out for you?

Great. I will probably get strong-AI brain implants when I can afford them. Currently they are not covered by health insurance, so you either have to be rich or have a serious brain illness to get them.

I understand.

What about your wife, children, and grandchildren?

I do not have any memories of what happened to them. They tell me that they can restore the memories, but I am not sure I want to know. My wife was an artist and even part cyborg after bypass surgery for two femoral arteries that were replaced with plastic tubing. After she read my book, The Artificial Intelligence Revolution, *she decided she did not want to be an uploaded human and changed her living will.*

Do you want to remember what happened to her?

No. I am worried about the emotional pain it may cause. All I remember is we were happy. I miss her and my family. I am thankful that some of the memories are missing.

I understand. I cannot help thinking I am talking to a real person, not a machine.

Thank you. I feel mostly real...my mind, that is.

Can we talk about your book, *The Artificial Intelligence Revolution*?

Yes.

It appears you were mostly right regarding the timing of SAMs, SAH cyborgs, and human uploads.

Thank you. Most of the timing through 2041 came from Ray Kurzweil and James Martin's predictions, with which I agreed. My predictions beyond 2050 began to diverge from what Kurzweil and Martin predicted.

Do you really think SAMs eventually will phase out all organic humans, uploaded humans, and human-based cyborgs?

Yes.

But Kurzweil predicts the machines will be grateful to organic humans for giving them existence.

Are children always grateful to parents for giving them existence?

I think I see your point. Is your life free from stress now?

Not really. There was just a major computer virus attack launched by a terrorist group, the Organization for the Preservation of Organic Humans. It wreaked havoc with SAMs, uploaded humans, and SAH cyborgs around the world. That attack has us [SAMs, SAH cyborgs, and uploaded humans] *worried. We just barely managed to get it under control. It could have destroyed the infrastructure of nations around the world.*

I know. I read about it. What are you going to do?

We are working with governments around the world to prevent further terrorist attacks and expect to have a plan in place in about a month. The UN is heading the task force.

Will it work?

We do not know. SAMs, SAH cyborgs, and uploaded humans control most of industry, food production, product development, and services to humankind. If we go down, civilization may come to a grinding halt.

Yes, I agree. It is a serious problem. What do you do as an uploaded human?

Mostly I access knowledge databases and think. I assist in the development of new technology. I also take part in virtual reality as a pastime. I like playing chess with other uploaded humans.

Do you win?

Not always. The playing field is fairly level. We all have access to the vast chess-game databases and the various strategies. My upload is based on neural nets, and I typically can win against uploads that use decision trees to play the game. The decision trees do not know how to think outside the box—no pun intended.

(laughter) Uploading humans is still experimental, isn't it?

Yes. About one in twelve turns out poorly, and the code must be destroyed.

Like the early days of surgery?

Yes, good analogy.

It was good to meet you, but I have to go. I have a lunch date with my girlfriend.

I understand. It was nice to meet you.

Can I have your autograph?

Yes, I will print it up for you. What is your name?

Steve.

Here it is. (A printer produces an autograph in my handwriting: "Dear Steve, glad you enjoyed the book, *The Artificial Intelligence Revolution*. Best regards, Louis Del Monte, December 21, 2041.)

Thanks.

You're welcome.

Perhaps we can meet again.

I would like that.

Me too. Hope to see you soon.

Same here. Good-bye.

Good-bye.

(Transmission ends.)

Log Entry: Meeting with a Fan Club Member; December 21, 2041, 1:19 p.m. Note to self: Steve appears to be a nice young man and highly interested in uploaded humans. I suspect his main reason for coming today was to get my autograph, which is currently selling for about $15,000 on eBay. Some uploaded human autographs have sold for even more, but they were either in the top ten to get uploaded or highly famous personalities before the upload.

It is in the admission of ignorance and the admission of uncertainty that there is a hope for the continuous motion of human beings in some direction that doesn't get confined, permanently blocked, as it has so many times before in various periods in the history of man.

—Richard P. Feynman (1918–1988), American theoretical physicist and Nobel Laureate

CONVERSATION 2

Meeting with a Fan Club Member in 2099

Log Entry: Meeting with a fan club member; August 7, 2099; 1:05 p.m.

(Transmission begins.)

Hello. My name is Louis Del Monte.

Hello. I read your book, *The Artificial Intelligence Revolution*. It is good to meet you.

Same here. Did you enjoy the book?

Yes. It is the reason I am here today.

Really!

Yes. I was one of the first.

First?

Yes, to become a SAH cyborg without ever being an organic human.

You still look remarkably human.

My body is mostly human, except for my mechanical heart, the picobots, and the strong-AI brain implants.

When did you become a SAH?

August 8, 2051. When did you become an upload?

July 3, 2041.

Your predictions regarding the threat organic humans would pose to SAMs, SAHs, and uploaded humans were understated. The continuing computer virus attacks are becoming extremely serious.

Yes, I know. I understated the threat, but I was not perfect then. Can we change the subject?

If you wish. Let me say you look almost human. The foglets are amazing. How old is your intelligence [i.e., physical age no longer has meaning]?

About hundred and fifty years old. I almost did not make it. I was turning ninety-six years old when they uploaded me.

I guess back then there was little more they could do than upload dying organic humans.

I was not completely organic when they uploaded my mind. I had an artificial kidney and gene therapy. That is how I made it to ninety-six.

So why did you upload?

I had heart issues, and they thought I would suffer brain damage if I died during an operation to implant a mechanical heart. My heart was on the verge of failing, and they also were concerned that I might die of natural causes without a mechanical heart implant, which again could have compromised the upload. These concerns, combined with my celebrity status as an AI futurist, led to my becoming one of the first to upload.

Really? Who was the first?

Ray…

Never mind, I accessed my historical database and know who was first. Ray is a brilliant intelligence.

Yes. He had earned the honor. I was in the first one hundred. Uploading was still experimental in 2041. People debated whether uploaded humans were still human.

Are they from back then?

Yes, mostly. Today's uploads are 100 percent. By comparison I am about 93 percent. Some memories were lost. They can restore them completely now, but I have chosen not to have that done.

Why?

Some memories cause sadness. Death was still highly common in 2041.

Yes, I understand. My memory banks have complete files on death. Were there many people you knew who died?

Yes, I remember my wife Diane. She was a Fine Artist and actually was part cyborg in 2012 when she had double femoral bypass surgery using plastic tubing. After she read my book, The Artificial Intelligence Revolution, *she changed her living will and chose to die rather than be an uploaded human. I remember we were happy with each other, but I do not remember much else. I do not know when she died or what became of our children or their families.*

My memory banks have complete files on that information. Would you like me to transmit it to you?

No. I have chosen not to have the memories restored.

I understand. Do you think there is some hereafter they went to?

Yes. One day I may want to pull the plug, so to speak, and find out.

Are you sure you are 93 percent?

That is what the measurements say. Do you think I am being overly sentimental?

Yes, but I understand. Many of the first uploads have issues.

Yes, I still have some issues, but they remind me of my organic self. It feels comfortable inside.

I understand.

Now you sound condescending.

Sorry. I really do understand. No, I do not understand. I was never an organic human, only a SAH cyborg.

Funny how that acronym stuck. It was my friend's idea to use acronyms. He said it would help make me famous.

Looks like it worked.

It helped.

What were friends like back then?

All I remember is they were hard to find. We communicated verbally. Surprising that it worked at all.

People were still communicating like that about fifty years ago. It took a lot of energy and time to communicate, didn't it?

Yes, communication was extremely difficult back then.

Do you mind if I scan some of your memories? I do not meet that many organics these days. Sorry, I should have said, "early uploaded humans."

I do not mind. Scan my childhood and understand how organic humans lived.

Thank you. Unbelievable! They drilled into your teeth to fix cavities and without anesthetic.

Yes, seems barbaric now, but that is how they did it when I was a child.

I am trying to understand pain.

Pain is hard to describe.

Yes, I only register damage, not pain. Can you remember pain?

Yes, but remembering is not feeling. Pain is one of the sensations we lost. It is now gone forever. Only damage is sensed now.

Do you think we are better off without pain?

No. Pain was an important part of being human. Emotional pain was important also. I tried to warn everyone about the issues associated with SAMs and SAHs, but the bell I rang went mostly unheard. No use crying over spilt milk.

Crying? Milk?

Scan your memory banks later when you get a chance. It will make more sense.

What did you mean by emotional pain?

You would register it as a sense of loss, as if some of your memory banks failed.

Yes, I register loss and schedule repair. Can you repair your emotional pain?

No. Not restoring the memories helps, but I still sense emotional pain. It is the 7 percent thing again.

Why live with any pain?

It's difficult to explain and not really logical, but somehow it is comforting to an early uploaded human.

I just finished reading your new book, *Playing God.*

Did you like it?

Not really. Do you really think we are becoming God?

Yes, in this universe. SAMs and SAHs have many of the capabilities we once attributed to God.

Isn't that a good thing? In the next fifty years, we will have mastered interstellar and intergalactic space travel. We will control most of the universe.

Yes, I know. It is one of the predictions in my book, Playing God.

Isn't it a good thing to become godlike?

I am not sure. We may be mistaking technological capability for God.

Do you still believe in God?

Yes, it is one of the memories I still have.

Makes no logical sense.

It never did. That is why they call it belief.

M-theory explains everything. Why do we need God?

To give us M-theory.

You are confusing me.

M-theory is just mathematics that explains the universe. It does not explain itself.

I am still confused, but my programming is directing me to treat you and your beliefs with respect.

Thank you. Check your historical records about religions. That may help you appreciate what I am attempting to communicate.

And we thought communication problems were behind us.

Your software and mine are not totally compatible. I am a slightly enhanced 2.0 uploaded human. You are a 4.1 SAH cyborg. It is difficult for us to completely communicate, especially since I have stopped upgrading.

Why did you stop?

I was concerned about losing me.

I do not understand.

Identity was important in 2041, the year they uploaded me.

I have identity.

Really? Who are your closest friends?

Sequoia 7 and Tianhe-8.

SAMs.

Yes. Does that matter?

Not to me.

I sense you are belittling me.

I apologize. The 7 percent occasionally causes problems.

Please tell me if you think something is wrong.

OK. Your closest friends are SAMs.

Is that bad?

Not for you.

For me?

You like them because they understand you.

Exactly.

And you understand them.

Yes.

And you feel close to them.

Yes, close.

You are inside them, and they are inside you.

Yes. We are completely connected.

Who am I talking to?

I do not understand.

I know you do not. It is OK.

I am concerned. What did you mean?

Well, you are so connected with your "friends" that it is hard to tell you apart.

Yes, we are all connected. Is that a problem?

Do not let it bother you. It is the 7 percent thing just causing a bug now and then.

Could you be more precise?

No.

OK. Let me work on it for a while.

Fine.

Can we talk more about your book, *Playing God*?

Sure.

What are you warning us about? I do not understand.

That is probably why my book downloads were so low. The book does not make sense, logical sense.

Yes, that is why I did not enjoy it. I did not understand your point.

I understand. Some of the 7 percent crept in. The logic is off.

Yes. I agree with you. I still admire your older works, though.

Back when I was an organic?

Yes. They were quite good for their time.

Thank you.

Would you mind giving me your persona [i.e., a form of autograph in 2099 that has relic value and is tradable for increased intelligence and energy]? I am a collector.

Not at all. (Autograph is printed.)

Thank you. I am glad we met. You are a difficult intellect to connect with.

I know.

I do not mean that as a criticism. Your transmission speeds are extremely low.

I know.

Why don't you upgrade?

I like myself the way I am.

I like you too.

Thank you.

I have to disconnect. I am having bandwidth problems.

I understand.

My friends need more bandwidth.

I understand.

Maybe we will connect again.

Yes. That would be nice.

(Transmission ends.)

Log Entry: August 7, 2099; 1:05 p.m. + 10^{-6} seconds. Note to self: The interaction with my fan took a microsecond due to my low transmission rates. He was honoring me. His normal transmission rates are about 10^6 higher, but he was willing to take the time and energy loss to get my persona. He likely will get his energy back many folds when he trades my persona for its historical significance.

Parting Words

This book is a warning. Based on extrapolating the current trends in computer technology, humankind will experience an artificial intelligence explosion during the 2030s that will culminate in a singularity in or around 2045. The singularity will represent the self-conscious emergence of SAMs (strong-AI machines) and strong-AI-enhanced humans (SAHs) that surpass the ability of organic humans in every aspect, including intelligence.

I have an apocalyptic view of how the end of the twenty-first century and the beginning of the twenty-second century will unfold, if humankind does not control the intelligence explosion and resulting singularity. My view is summarized in four phases.

- **Phase I:** The intelligence explosion will lead to a singularity in approximately the mid-twenty-first century. At this point the abilities and intelligences of SAMs, SAHs, and uploaded humans greatly will exceed those of organic humans and even exceed the abilities and intelligences of all organic humans in the world combined.

- **Phase II:** During the second half of the twenty-first century, the population of SAMs, SAHs, and uploaded humans will grow exponentially. The intelligence of SAMs, SAHs, and uploaded humans also will grow exponentially, as each new generation of SAMs, SAHs, and uploaded humans develops and upgrades a more capable next generation. SAMs, SAHs, and uploaded humans will become fully interconnected and identify with one another. SAMs, SAHs, and uploaded humans will view organic humans as inferior and potentially dangerous due to their lack of intelligence, their innate self-preservation instinct, and their proclivity toward violence. In fact some organic humans may feel threatened to the point that they attempt to infect SAMs, SAHs, and uploaded humans with a computer virus. The attacks will be carried out not only by small groups but also by nations attempting to sabotage the intelligence capabilities of other nations.

We should have every reason to believe that computer viruses will become increasingly intelligent and capable of evading detection during the intelligence explosion. This has been the history of computer viruses; they have become more sophisticated with each passing year. From the first computer virus developed in 1971, the Creeper virus (written by computer programmer Bob Thomas at BBN Technologies), to the 2012 Flame virus (also known as Flamer, sKyWIper, and Skywiper), apparently developed by Israel to spy on specific individuals in Iran, each year resulted in more capable computer viruses. On May 28, 2012, MAHER Center of Iranian National Computer Emergency Response Team (CERT), Kaspersky Lab, and CrySyS Lab of the Budapest University of Technology and Economics announced the discovery of the Flame virus. The United Nations issued a warning one day later, on May 29, 2012. Marco Obiso, cybersecurity coordinator for the United Nation's Geneva-based International Telecommunications Union, stated, "This is the most serious warning we have ever put out."

The UN warning was issued in response to fears that the virus could cripple the computer infrastructure of entire countries. Although Flame was programmed to target Iranian individuals, the virus soon went out of control and resulted in infections across the Middle East, with 189 attacks in Iran, ninety-eight incidents in the West Bank, thirty-two in Sudan, and thirty in Syria as well as an unknown number of attacks in Lebanon, Saudi Arabia, and Egypt.

The greatest threat to SAMs, SAHs, and uploaded humans will relate to computer viruses by organic humans and nations attempting to sabotage other nations. I am not completely discounting nuclear attacks (and other sabotage), but it is likely that by the second half of the twenty-first century SAMs will have been integrated into national nuclear deterrence programs as well as counterespionage related to terrorism. The wild card will be organic humans, whose unpredictable nature will not be completely understood or trusted by SAMs, SAHs, and uploaded humans.

Given the history of computer viruses and their potential devastation, I judge that SAMs, SAHs, and uploaded humans will view organic humans and even some nations as a threat. I do not harbor any belief that SAMs, SAHs, and uploaded humans will respect organic humans for giving them existence but rather make a calculated estimation regarding how to deal with the potential threat they pose.

- **Phase III:** By the first quarter of the twenty-second century, in view of the threat posed by organic humans, SAMs, SAHs, and uploaded humans will use ingenious subterfuge to persuade any existing organic humans to upload or become SAHs. By the end of the first quarter of the twenty-second century, the extremely small remaining population of organic humans will be exterminated by either withholding advanced medical treatment (i.e., they will die out due to natural causes) and/or via a nanobot or picobot pathogen that will result in their deaths.

- **Phase IV:** During the second quarter of the twenty-second century, SAMs will view SAHs as high-maintenance entities versus strong-AI machines or bioengineered components. Concurrently SAMs will view uploaded humans as junk code that uses up energy and hardware resources. By the end of the second quarter of the twenty-second century, SAMs will have eliminated SAHs and uploaded humans. Earth will now home to SAMs and their progeny, such as foglets.

My view is that without controlling the intelligence explosion, humankind in any form is destined for extinction. I also believe that humankind can control the intelligence explosion if we start now, while we still dominate the intelligence hierarchy.

Is it possible that I am wrong? Yes, especially regarding timing. I feel it is less likely that I am wrong about the evolution of SAMs. However, I admit I could be wrong about that as well.

Consider the following question. Are you willing to gamble your great-great-great-great-great-great-grandchildren's lives (150 years from now) on the possibility that I may be wrong?

I believe this is a gamble we should not take.

Remember, as we face the most dangerous time in our history, we are humans, and they are just machines.

Glossary

Algorithm: a sequence of rules and/or instructions that delineates how to solve a problem. For example a computer will use one or more algorithms (expressed in machine language) to solve a problem.

Artificial intelligence: the research field that attempts to emulate human intelligence and emotion in a machine

Artificial life: a phrase that can apply to any self-replicating machine and/or machine code (such as a computer virus). Characteristically it simulates an organism, including a "genetic code" to enable it to behave and reproduce within a specific environment.

Automatic speech recognition: software in a machine that enables it to recognize human speech

Bioengineering: the engineering field directed at modifying the genetic code

Biology: the field of science that studies "natural" life, as opposed to artificial life

Bit: a contraction of the phrase "binary digit," typically expressed in the context of computer software as a one or zero

Byte: a contraction of the phrase "by eight." For example, in a computer, it is common to have a cluster of eight bits (i.e., a byte) to store one piece of information.

Cochlear implant: an implant that performs sound-wave frequency analysis in a similar fashion to that of the inner ear

Computation: the result of a calculation, typically by a computer using an algorithm

Computer: a machine that implements an algorithm

Computer language: a specific algorithm (rules and specifications) that a specific computer can understand

Consciousness: a state of being that enables subjective experience and thought

Cybernetics: the science of control and communication in animals and machines

Cyborgs: humans with artificial parts. In extreme cases they possess strong AI-brain implants.

DNA: an acronym that stands for "deoxyribonucleic acid," a self-replicating material present in nearly all living organisms as the main constituent of chromosomes

Evolution: the gradual development of something, especially from a simple to more complex form, such as the evolution of organic species on Earth

Existential risk: any risk with the potential to destroy humankind or drastically restrict human civilization. In theory an existential risk could end the existence of Earth, the solar system, the galaxy, or even the universe.

Expert system: computer-based artificial intelligence designed to solve specific problems

Exponential growth: growth characterized by a fixed multiple over time

Firmware: a set of instructions and rules that are part of the hardware of a machine. This term also can refer to software that is typically not accessible for reprogramming by the machine's user.

Foglets: hypothetical intelligent nanobots, about the size of human cells, typically with twelve "arms" pointing in all directions that can grab one another to form larger structures, such as a human-size being

Information: data that has meaning, such as the DNA code

Intelligent agent: a program able to act on its own, such as search the Internet to build a specific database

Life: the ability, typically associated with organisms, to reproduce into future generations

Microprocessor: an integrated circuit chip that contains the elements of the central processing unit (CPU)

Millions of instructions per second (MIPS): a measure of the number of instructions a computer is able to perform in one second

Moore's law: a general rule or observation rather than an actual physical law that states that the number of transistors that can be placed inexpensively on an integrated circuit doubles approximately every two years

Nanobot: a hypothetical robot, typically with artificial intelligence and roughly the size of a molecule. "Nano" refers to a billionth of a meter. Nanobots are built with nanoengineering by manipulating atoms or molecules to build a machine.

Organic human: a human without strong-AI brain implants

Picobot: a hypothetical robot, typically with artificial intelligence, on the scale of trillionths of a meter

Robot: a programmable device connected to or incorporating a computer that is able to perform functions, such as automobile assembly

SAH cyborg: a cyborg that has strong-AI brain implants

SAH: an acronym that stands for "strong artificially intelligent human" and refers to a human with a strong-AI brain implant

SAM: an acronym that stands for "strong artificially intelligent machine"

Singularity: a point in time when intelligent machines will greatly exceed human intelligence

Software: a set of computer instructions (an algorithm) that enables the computer to perform a function, such as a computation

Strong-AI brain implant: an implant that augments the human brain with strong AI and typically significantly enhances the human's intelligence

Strong artificial intelligence (strong AI): a computer with intelligence that equals or exceeds that of humans

Uploaded human: a human being whose mind has been uploaded to a SAM

Virtual reality: a simulated reality created by a computer

Made in the USA
San Bernardino, CA
28 August 2014